# Frequency-Domain Receiver Design for Doubly Selective Channels

# Frequency-Domain Receiver Design for Doubly Selective Channels

Paulo Montezuma
Fabio Silva
Rui Dinis

## CRC Press

Taylor & Francis Group

Boca Raton   London   New York

CRC Press is an imprint of the
Taylor & Francis Group, an **informa** business

CRC Press
Taylor & Francis Group
6000 Broken Sound Parkway NW, Suite 300
Boca Raton, FL 33487-2742

First issued in paperback 2019

© 2017 by Taylor & Francis Group, LLC
CRC Press is an imprint of Taylor & Francis Group, an Informa business

No claim to original U.S. Government works

ISBN-13: 978-1-138-70092-5 (hbk)
ISBN-13: 978-0-367-88841-1 (pbk)

**Visit the Taylor & Francis Web site at**
**http://www.taylorandfrancis.com**

**and the CRC Press Web site at**
**http://www.crcpress.com**

# Contents

# Authors

**Paulo Montezuma** received a PhD from Faculdade de Ciências e Tecnologia, Universidade Nova de Lisboa (FCT-UNL), Portugal, in 2007. Currently, he is professor at FCT-UNL. From 1992 to 2007, he was a researcher at CAPS (Centro de Análise e Processamento de Sinal), IST. Since 2001, he has been a researcher at Uninova, and since 2013 he has also been a researcher at IT (Instituto de Telecomunicações). Dr. Montezuma has been actively involved in several national  and international research projects in the broadband wireless communications area. His research interests include: coding, wireless transmission, estimation, and detection techniques.

**Fabio Silva** received a PhD from Faculdade de Ciências e Tecnologia, Universidade Nova de Lisboa (FCT-UNL), Portugal, in 2015. Currently, he is an engineer at Anacom, a Portuguese telecommunications regulator. He was a researcher at Uninova-FCT from 2008 to 2015. Since 2013, he has been a researcher at IT (Instituto de Telecomunicações). From 2010 to 2015 Dr. Silva was actively involved in several national research  projects in the broadband wireless communications area. His research interests include: radio transmission, channel estimation, and detection techniques.

**Rui Dinis** received a PhD from Instituto Superior Técnico (IST), Technical University of Lisbon, Portugal, in 2001, and the Habilitation in Telecommunications from Faculdade de Ciências e Tecnologia (FCT), Universidade Nova de Lisboa (UNL), in 2010. From 2001 to 2008 he was a professor at IST. Currently, he is an associate professor at FCT-UNL. During 2003, he was an invited professor at Carleton University, Ottawa, Canada. Dr. Dinis was a researcher at CAPS (Centro de Análise e Processamento de Sinal), IST, from 1992 to 2005, and a re-  searcher at ISR (Instituto de Sistemas e Robótica) from 2005 to 2008. Since 2009, he has been a researcher at IT (Instituto de Telecomunicações). He has been actively involved in several national and international research projects in the broadband wireless communications area. His research interests include: transmission, estimation, and detection techniques. Dr. Dinis is editor of *IEEE Transactions on Communications* (Transmission Systems - Frequency-Domain Processing and Equalization) and *IEEE Transactions on Vehicular Technology*. He was also a guest editor for Elsevier's *Physical Communication* (Special Issue on Broadband Single-Carrier Transmission Techniques).

# List of Abbreviations

**ADC** Average Doppler Compensation

**AWGN** Additive White Gaussian Noise

**BER** Bit Error Rate

**BLUE** Best Linear Unbiased Estimator

**CFO** Carrier Frequency Offset

**CP** Cyclic Prefix

**CIR** Channel Impulse Response

**CLT** Central Limit Theorem

**DAB** Digital Audio Broadcasting

**DAC** Digital-to-Analog Converter

**DFT** Discrete Fourier Transform

**DFE** Decision Feedback Equalizer

**DVB** Digital Video Broadcasting

**DVB-T** Digital Video Broadcasting - Terrestrial

**EIRP** Effective Isotropic Radiated Power

**FDE** Frequency-Domain Equalization

**FDM** Frequency Division Multiplexing

**FFT** Fast Fourier Transform

**FIR** Finite Impulse Response

**IB-DFE** Iterative Block-Decision Feedback Equalizer

**IDFT** Inverse Discrete Fourier Transform

**IFFT** Inverse Fast Fourier Transform

**IBI** Inter-Block Interference

**ICI** Inter-Carrier Interference

**ISI** Inter-Symbol Interference

**LLR** Log-Likelihood Ratio

**LOS** Line-Of-Sight

**MC** Multicarrier

**MFB** Matched Filter Bound

**MLR** Maximum Likelihood Receiver

**MMSE** Minimum Mean Square Error

**MRC** Maximal-Ratio Combining

**MSE** Mean Square Error

**OFDM** Orthogonal Frequency-Division Multiplexing

**PAPR** Peak-to-Average Power Ratio

**PMEPR** Peak-to-Mean Envelope Power Ratio

**PDF** Probability Density Function

**PDP** Power Delay Profile

**PSD** Power Spectral Density

**PSK** Phase Shift Keying

**QAM** Quadrature Amplitude Modulation

**QPSK** quadrature Phase-Shift Keying

**SC** Single Carrier

**SC-FDE** Single Carrier with Frequency Domain Equalization

**SINR** Signal to Interference-plus-Noise Ratio

**SFN** Single Frequency Networks

**SNR** Signal to Noise Ratio

**SISO** Soft-In, Soft-Out

**TDC** Total Doppler Compensation

**ZF** Zero-Forcing

# List of Symbols

## General Symbols

| | |
|---|---|
| $A_e$ | effective area or "aperture" of an antenna |
| $B$ | bandwidth of a given frequency domain signal |
| $B_D$ | Doppler spread |
| $B_k$ | feedback equalizer coefficient for the $k^{th}$ frequency |
| $B_C$ | coherence bandwidth |
| $c$ | speed of light (in $m/s$) |
| $c_l(\tau, t)$ | channel response, at time $t$, to a pulse applied at $t - \tau$ |
| $E_b$ | average bit energy |
| $E_s$ | average symbol energy |
| $F$ | subcarrier separation |
| $F_k$ | feedforward equalizer coefficient for the $k^{th}$ frequency |
| $F_k^{(l)}$ | feedforward equalizer coefficient for the $k^{th}$ frequency and $l^{th}$ diversity branch |
| $f$ | frequency variable |
| $f_c$ | carrier frequency |
| $f_D$ | maximum Doppler frequency |
| $f_D^{(r)}$ | Doppler drift associated with the $r^{th}$ cluster of rays |
| $f_k$ | $k^{th}$ frequency |
| $f_0$ | fundamental frequency |
| $G_t$ | gain of the transmitter antenna |
| $G_r$ | gain of the receiver antenna |
| $g(t)$ | impulse response of the transmit filter |
| $i$ | tap index of the diversity branch |
| $H_k$ | overall channel frequency response for the $k^{th}$ frequency |
| $\tilde{H}_{kL}$ | overall channel frequency response estimation for the $k^{th}$ frequency |
| $H_k^{(l)}$ | overall channel frequency response for the $k^{th}$ frequency and $l^{th}$ |

|  | diversity branch |
|---|---|
| $H_k^{(m)}$ | overall channel frequency response for the $k^{th}$ frequency of the $m^{th}$ time block |
| $\tilde{H}_k^D$ | data overall channel basic frequency response estimation for the $k^{th}$ frequency |
| $\hat{H}_k^D$ | data overall channel enhanced frequency response estimation for the $k^{th}$ frequency |
| $\tilde{H}_k^{TS}$ | training sequence overall channel basic frequency response estimation for the $k^{th}$ frequency |
| $\hat{H}_k^{TS}$ | training sequence overall channel enhanced frequency response estimation for the $k^{th}$ frequency |
| $\tilde{H}_k^{TS,D}$ | overall channel basic frequency response combined estimation for the $k^{th}$ frequency |
| $\hat{H}_k^{TS,D}$ | overall channel enhanced frequency response combined estimation for the $k^{th}$ frequency |
| $h(t)$ | channel impulse response |
| $h_b(t)$ | complex baseband representation of $h(t)$ |
| $h_T(t)$ | pulse shaping filter |
| $\tilde{h}_n^D$ | data overall channel basic impulsive response estimation for the $n^{th}$ time-domain sample |
| $\hat{h}_n^D$ | data overall channel enhanced impulsive response estimation for the $n^{th}$ time-domain sample |
| $\tilde{h}_n^{TS}$ | training sequence overall channel basic impulsive response estimation for the $n^{th}$ time-domain sample |
| $\hat{h}_n^{TS}$ | training sequence overall channel enhanced impulsive response estimation for the $n^{th}$ time-domain sample |
| $\tilde{h}_n^{TS,D}$ | overall channel basic impulsive response combined estimation for the $n^{th}$ time-domain sample |
| $\hat{h}_n^{TS,D}$ | overall channel enhanced impulsive response combined estimation for the $n^{th}$ time-domain sample |
| $J_0$ | zeroth-order Bessel function of the first kind |
| $k$ | frequency index |
| $L$ | number of paths within a multipath fading channel |
| $L_s$ | system losses due to hardware |
| $L_n^{I(i)}$ | in-phase log-likelihood ratio for the $n^{th}$ symbol at the $i^{th}$ iteration |
| $L_n^{Q(i)}$ | quadrature log-likelihood ratio for the $n^{th}$ symbol at the $i^{th}$ |
| $l$ | iteration antenna index/diversity branch |
| $m$ | data symbol index |
| $N$ | number of symbols/subcarriers |
| $N_0$ | noise power spectral density (unilateral) |
| $N_D$ | number of data blocks |
| $N_{Rx}$ | space diversity order |
| $N_{TS}$ | number of symbols of the training sequence |

| | |
|---|---|
| $N_k$ | channel noise for the $k^{th}$ frequency |
| $N_k^{TS}$ | training sequence channel noise for the $k^{th}$ frequency |
| $N_k^{(l)}$ | channel noise for the $k^{th}$ frequency and $l^{th}$ diversity branch |
| $N_k^{(m)}$ | channel noise for the $k^{th}$ frequency of the $m^{th}$ time block |
| $N_G$ | number of guard samples |
| $n$ | time-domain sample index |
| $n(t)$ | noise signal |
| $PL$ | *path-loss* for the free space model |
| $\overline{PL}$ | mean *path-loss* in dB |
| $P_b$ | AWGN channel performance |
| $P_{b,\text{MFB}}$ | matched filter bound performance |
| $P_{b,\text{Ray}}$ | performance of a single ray transmitted between the transmitter and the receiver |
| $P_e$ | bit error rate |
| $\hat{P}_e$ | estimated bit error rate |
| $P_r(d)$ | received power as a function of the distance $d$ |
| $P_t$ | transmitted power |
| $P_t$ | transmitted power |
| $R$ | symbol rate |
| $R(f)$ | Fourier transform of $r(t)$ |
| $R(\tau)$ | autocorrelation function |
| $r(t)$ | rectangular pulse/shaping pulse |
| $r_{h_b h_b}()$ | autocorrelation function of $h_b(t)$ |
| $S(f)$ | frequency-domain signal |
| $S_k$ | $k^{th}$ frequency-domain data symbol |
| $S_k^{TS}$ | training sequence $k^{th}$ frequency-domain data symbol |
| $S_k^{(m)}$ | $k^{th}$ frequency-domain data symbol of the $m^{th}$ data block |
| $\tilde{S}_k$ | estimate for the $k^{th}$ frequency-domain data symbol |
| $\tilde{S}_k^{(m)}$ | estimate for the $k^{th}$ frequency-domain data symbol of the $m^{th}$ data block |
| $\hat{S}_k$ | "hard decision" for the $k^{th}$ frequency-domain data symbol |
| $\hat{S}_k^{(m)}$ | "hard decision" for the $k^{th}$ frequency-domain data symbol of the $m^{th}$ data block |
| $\overline{S}_k$ | "soft decision" for the $k^{th}$ frequency-domain data symbol |
| $s(t)$ | time-domain transmitted signal |
| $s_b(t)$ | complex baseband representation of $s(t)$ |
| $s(t)^{(m)}$ | signal associated with the $m^{th}$ data block |
| $s^I(t)$ | continuous in-phase component |
| $s^Q(t)$ | continuous quadrature component |
| $s_n$ | $n^{th}$ time-domain data symbol |
| $s_n^I$ | discrete in-phase component |
| $s_n^Q$ | discrete quadrature component |
| $s_n^{TS}$ | training sequence $n^{th}$ symbol |
| $s^\Delta$ | time-domain transmitted signal affected by carrier frequency |

# Greek Letters Symbols

$\alpha_l$      attenuation of given multipath component

$\beta$      relation between the average power of the training sequences and the data power

$\Delta_f$      carrier frequency offset

$\Delta_k^{(i)}$      error term for the $k^{th}$ frequency-domain "hard decision" estimate

$\Delta_k^{(m)}$      zero-mean error term for the $k^{th}$ frequency-domain "hard decision" estimate of the $m^{th}$ data block

$\gamma^{(i)}$      average overall channel frequency response at the $i^{th}$ iteration

$\kappa^{(i)}$      normalization constant for the FDE

$\lambda_c$      wavelength of the carrier frequency (measured in meters)

$\rho^{(i)}$      correlation coefficient at the $i^{th}$ iteration

$\rho_m$      correlation coefficient of the $m^{th}$ data block

$\rho_n^I$      correlation coefficient of the "in-phase bit" of the $n^{th}$ data symbol

$\rho_n^Q$      correlation coefficient of the "quadrature bit" of the $n^{th}$ data symbol

$\sigma$      standard deviation

$\sigma_{Eq}^2$      total variance of the overall noise plus residual ISI

$\hat{\sigma}_{Eq}^2$      approximated value of $\sigma_{Eq}^2$

$\sigma_{MSE}^2$      mean-squared error (MSE) variance

$\sigma_N^2$      variance of channel noise

$\sigma_S^2$      variance of the transmitted frequency-domain data symbols

$\sigma_{H,TS}^2$      variance of the noise in the channel estimates related with the training sequence

$\sigma_D^2$      variance of the noise in the channel estimates related with the data blocks

$\sigma_T$      total received power from the scatterers affecting the channel at given delay $\tau$

$\sigma_{TS,D}^2$      variance of the noise in the combined channel estimates

$\Theta_k$      overall error for the $k^{th}$ frequency-domain sample

$\Theta(k)$      mean-squared error (MSE) in the time-domain

$\theta_l$      angle between the direction of the movement and the direction of departure of the $l^{th}$ component

$\theta_n$      phase rotation due to CFO associated with the $n^{th}$ sample

$\theta_n^{(r)}$      estimated phase rotation due to Doppler frequency drift

$\varepsilon_k^{(i)}$      global error consisting of the residual ISI plus the channel noise at the $i^{th}$ iteration

$\varepsilon_k^{Eq(i)}$      denotes the overall error for the $k^{th}$ frequency-domain symbol

$\vartheta_n^I$      error in $\hat{s}_n^I$

$\vartheta_n^Q$      error in $\hat{s}_n^Q$

$\Omega_{i,l}^2$      mean square value of the magnitude of each tap $i$ for the $l^{th}$

|  | diversity branch |
| $\omega_c$ | frequency carrier (in rads/s) |
| $\Phi_r$ | set of all multipath components |
| $\phi$ | phase offset |
| $\phi_{Dop,l}$ | Doppler phase shift of the $l^{th}$ multipath component |
| $\varphi_{i,l}(t)$ | zero-mean complex Gaussian random process |
| $\tau_{i,l}$ | delay associated with the $i^{th}$ tap and $l^{th}$ diversity branch |
| $\delta(t)$ | Dirac function |
| $\nu_l$ | represents AWGN samples |
| $\epsilon_{kL}^{H}$ | channel estimation error |
| $\epsilon_{k}^{D}$ | data channel estimation error |
| $\epsilon_{k}^{TS}$ | training sequence channel estimation error |
| $\epsilon_{k}^{TS,D}$ | combined training and data channel estimation error |

# Matrix and Arrays Symbols

| $\mathbf{z}$ | $U_{total} \times 1$ vector |
| $\mathbf{z}^{H}$ | conjugate transpose of $\mathbf{z}$ |

# Chapter 1

# Introduction

## 1.1 Motivation and Scope

The tremendous growth of mobile internet and multimedia services, accompanied by the advances in micro-electronic circuits as well as the increasing demands for high data rates and high mobility, motivated the rapid development of broadband wireless systems over the past decade. Future wireless systems are expected to be able to deploy very high data rates of services within high mobility scenarios. As a result, broadband wireless communication is nowadays a fundamental part of the global information and the world's communication structure.

A major challenge in the design of mobile communications systems is to overcome the mobile radio channel effects, assuring at the same time high power and spectral efficiencies. Since in mobile communications the information data is transmitted across the wireless medium, then the transmitted signal will certainly suffer from adverse effects originated by two different factors: multipath fading and mobility.

Within a multipath propagation environment waves arriving from different paths with different delays combine at the receiver with different attenuations. Multipath propagation leads to the time dispersion of the transmitted symbol resulting in **frequency-selective fading**.

Besides multipath propagation, time variations within the channel may also arise due to oscillator drifts, as well as due to mobility between transmitter and receiver [JCWY10]. The relative motion between the transmitter and the receiver results in Doppler frequency which has a strong negative impact on the performance of mobile radio communication systems since it generates different frequency shifts for each incident plane wave, causing the channel impulse response to vary in time. The channel characteristics change depending on the location of the user, and because of mobility, they also vary in time. Hence, when the relative positions of the different objects in the environment including the transmitter and receiver change with time, the nature of the channel also varies. In mobility scenarios, the rate of variation of the channel response in time is characterized by the *Doppler spread*. Significant variations of the channel response within the signal duration lead to **time-selective fading**, and this represents a major issue in wireless communication systems.

Channels whose response is selective in time and frequency are referred to as doubly-selective. As a result of these two phenomena, the equivalent received signal is time varying and may be highly attenuated. This is considered a severe impairment in wireless communication systems, since these effects lead to drastic and unpredictable fluctuations of the envelope of the received signal (deep fades of more than 40 dB below the mean value can occur several times per second).

Block transmission techniques, with cyclic extensions and FDE techniques (frequency-domain equalization) are known to be suitable for high data rate transmission over severely time-dispersive channels due to its reduced complexity and excellent performance, provided that accurate channel estimates are provided. Moreover, since these techniques usually employ large blocks, the channel can even change within the block duration. Fourth generation broadband wireless systems employ CP-assisted (cyclic prefix) block transmission techniques, and although these techniques allow the simplification of the receiver design, the length of the CP should be a small fraction of the overall block length, meaning that long blocks are susceptible to time-varying channels, especially for mobile systems. Hence, the receiver design for doubly-selective channels is of key importance, especially to reduce the relative weight of the CP.

Efficient channel estimation techniques are crucial in achieving reliable communication in wireless communication systems. When the channel changes within the block duration, significant performance degradation occurs. Channel variations lead to two different difficulties: first, the receiver needs continuously accurate channel estimates; second, conventional receiver designs for block transmission techniques are not suitable when there are channel variations within a given block. As with any coherent receiver, accurate channel estimation is mandatory for the good performance of FDE receivers, both for orthogonal frequency-division multiplexing (OFDM) and single carrier with frequency domain equalization (SC-FDE).

The existence of residual carrier frequency offset (CFO) between the transmitter and the receiver's local oscillators means that the equivalent channel has a phase rotation that changes within the block. It was shown in [SF08, AD04, DAPN10] that residual CFO leads to simple phase variations that are relatively easy to compensate at the receiver's side. However, that may not be the case for single frequency broadcast networks. Within single frequency networks (SFN), several receiving zones within the overall coverage location are served by more than one transmitter, meaning that multiple transmitters must broadcast the same signal simultaneously over the network. Hence, each transmission will most likely have an associated frequency offset. This leads to a very difficult scenario where there will be substantial variations on the equivalent channel which cannot be treated as simple phase variations.

For channel variations due to Doppler effects, receiver structures for double-selective channels combining an iterative equalization and compensation of channel variations have already been proposed [NF04]. These kinds of channel variations can become extremely complex since the Doppler effects are distinct for different multipath components (e.g., when we have different departure/arrival directions relatively to the terminal movement).

It is difficult to ensure stationarity of the channel within the block duration, which is a requirement for conventional OFDM and SC-FDE receivers. Hence, efficient estimation and tracking procedures are required and should be able to cope with channel variations.

This book is dedicated to the study of effective detection of broadband wireless transmission, and it is intended for future broadband wireless and cellular systems which should be able to provide high transmission, together with high mobility (e.g., WiFi/WiMax-type LANs). Contrary to the common approach that assumes that either the channel is fixed or non-dispersive, this book focuses on the problem of digital transmission over severely time-dispersive channels that are also time-varying. Both OFDM and SC-FDE schemes will be considered. Effective detection within channels that are both time dispersive and time varying can be achieved by resorting to receiver designs implemented in the frequency domain, capable of performing channel estimation and compensation, as well as channel tracking techniques. Therefore this book aims to present these techniques for estimating the channel impulse response and track its variations. These techniques should take advantage of reference symbols/block multiplexed with data and/or added to it. Finally are presented receivers analyzed for severely time-dispersive channels that combine the detection/equalization procedures with the channel estimation techniques, while assuring low and moderate signal processing requirements.

## 1.2 Book Structure

Chapter 2 is devoted to a review of the mathematical models representing the physical channels and introduces time-varying frequency selective channels. Having in mind the high data rate requirements while dealing with severely time-dispersive channel effects, this chapter includes an overview of the state-of-the-art of the equalization techniques at the receiver side that become necessary to compensate the signal distortion and guarantee good performance. Channel characterization covers both frequency and time selective channels. So, besides multipath propagation resulting in frequency-selective fading, channels with time variations within the channel may also arise due to oscillator drifts, as well as due to motion between transmitter and receiver.

This characterization also covers channels whose response is selective in time and frequency. These are referred to as doubly-selective and represent a severe impairment in wireless communication systems, since the multipath propagation combined with the Doppler effects due to mobility can lead to drastic and unpredictable fluctuations of the envelope of the received signal.

Chapter 3 starts with a brief introduction of OFDM and SC-FDE block transmission techniques that are especially adequate for severely time-dispersive channels. It includes several aspects such as the analytical characterization of each modulation type and some relevant properties of each modulation. For both modulations special attention is given to the characterization of the transmission and receiving structures, with particular emphasis on transmitter and receiver performance. MC modulations and their relations with SC modulations are analyzed. Section 3.4.1 describes the OFDM modulation. Section 3.5 characterizes the basic aspects of the SC-FDE modulation including the linear and iterative FDE receivers. A promising FDE technique for single carrier modulation, the iterative block-decision feedback equalizer (IB-DFE), is also analyzed and an explanation of the feedforward and the feedback operations is given. It is shown that this iterative FDE receiver offers much better performance than the non-iterative methods, with performance near to the MFB as will be shown in Chapter 4. Finally, in Section 3.6, the performance of OFDM and SC-FDE for severely time-dispersive channels is compared.

In Chapter 4, the impact of the number of multipath components and the diversity order on the asymptotic performance of OFDM and SC-FDE for different channel coding schemes is analyzed. It is shown that the number of relevant separable multipath components is a fundamental element that influences the performance of both schemes and, in the IB-DFE's case, the iteration gains. A set of results is presented that demonstrates that SC-FDE has an overall performance advantage over OFDM, especially when employing the IB-DFE, in the presence of a high number of separable multipath components, because it allows a performance very close to the matched filter bound (MFB), even without diversity. With diversity, the performance approaches MFB faster, even for a small number of separable multipath components.

Chapter 5 is devoted to OFDM-based broadcasting systems with SFN operation and presents an efficient channel estimation method which takes advantage of the sparse nature of the equivalent channel impulse response (CIR). For this purpose, low-power training sequences are used in order to obtain an initial coarse channel estimate, and an iterative receiver with joint detection and channel estimation is designed. The results achieved by this receiver show very good performance, close to the perfect channel estimation case, even with resort to low-power training blocks as well as for the case where the receiver does not know the location of the different clusters that constitute the overall CIR.

Chapter 6 is dedicated to the joint CFO estimation and compensation over the severe time-distortion effects inherent in SFN systems. Most conventional broadband broadcast wireless systems employ OFDM schemes in order to cope with severely time-dispersive channels. As shown in Chapter 3, the high peak-to-average power ratio (PAPR) of OFDM signals leads to amplification difficulties. Moreover, the presence of a carrier frequency offset compromises the orthogonality between the OFDM subcarriers. Thus, this chapter explores the possibility of using SC-FDE schemes in broadcasting systems with SFN operation. An efficient method for estimating the channel frequency response and CFO associated with each transmitter is presented, along with receiver structures able to compensate the equivalent channel variations due to different CFO for different transmitters. Subsequently, an efficient technique is also presented for estimating the channel associated with the transmission between each transmitter and the receiver, as well as the corresponding CFOs. This technique has been shown to be sufficient for obtaining the evolution of the equivalent channel along a given frame. Closing this chapter, it is analyzed a set of iterative FDE receivers able to compensate the impact of the different CFOs between the local oscillators at each transmitter is analyzed.

Finally, Chapter 7 focuses on the problem of the use of SC-FDE transmission in channels with strong Doppler effects. For this purpose, new iterative frequency-domain receivers able to attenuate the impact of strong Doppler effects, at the cost of a slight increase in complexity when compared with the IB-DFE, are defined. The first step is to do a channel characterization appropriate to model short-term channel variations, modeled as almost pure Doppler shifts which were different for each multipath component. Then, this model will be used to design the frequency-domain receivers able to deal with strong Doppler effects. These receivers can be considered as modified turbo equalizers implemented in the frequency-domain, which are able to compensate the Doppler effects associated with different groups of multipath components while performing the equalization operation, which makes them suitable for SC-FDE scheme based broadband transmission in the presence of fast-varying channels.

# Chapter 2

# Fading

In order to enable communication over wireless channels it is necessary to characterize the propagation models. However, trying to make an analysis of the mobile communication under such harsh propagation conditions might seem a very hard task to accomplish. Nevertheless, starting from a model based on the multipath propagation we will see that many of the properties of the transmission can be successfully predicted by applying powerful techniques of statistical communication theory [JC94].

One of the major challenges in the design of mobile communications systems is to overcome the effects of mobile radio channels, assuring at the same time reliable high-speed communication. Parameters like the paths taken by the multipath components, the presence of objects along these paths, and the distance between the transmitter and receiver have a direct influence on the signal.

The wireless channel experiences deep fade in time or frequency. Fading effects related to mobile communications can be classified in two spatial scales:

- Large-scale fading: based on *path-loss* and shadowing;

- Small-scale fading: based on multipath fading and Doppler spread.

## 2.1  Large-Scale Fading

As the name suggests, large-scale fading refers to variations in received power over large distances. In this section, we characterize these variations in received signal power over distance, which are due to *path-loss* and *shadowing*.

7

### 2.1.1 Path-Loss

The signal attenuation of an electromagnetic wave (represented by a reduction in its power density), between a transmitting and a receiving antenna as a function of the propagation distance, is called *path-loss*. As the relative distance between the transmitter and receiver increases, the power radiated by the transmitter dissipates as the radio waves propagate through the channel. This is commonly referred to as *free-space path-loss* and refers to a signal propagating between the transmitter and receiver with no attenuation or reflection. This is the simplest model for signal propagation and is based on the free-space propagation law. Let us consider the free-space propagation model. It considers the line of sight channel in which there are no objects between the receiver and the transmitter, and it attempts to predict the received signal strength assuming that power decays as a function of the distance between the transmitter and receiver.

The Friis free space equation states that for a transmission between a transmitter and receiver separated by a distance $d$, the power acquired by the receiver's antenna, as a function of the $d$, is given by [Rap01],

$$P_r(d) = \frac{P_t G_t G_r \lambda^2}{(4\pi)^2 d^2 L_s},\tag{2.1}$$

where $P_t$ stands for the transmitted power (assumed to be known in advance), $G_t$ and $G_r$ represent the gains at the transmitter and receiver antenna, respectively, considering that both antennas are isotropic. The parameter $\lambda$ is the wavelength measured in meters, and is related to the carrier frequency by

$$\lambda = \frac{c}{f_c} = \frac{c}{2\pi/\omega_c},\tag{2.2}$$

with $c$ representing the speed of light (in m/s), $f_c$ representing the carrier frequency (in Hertz), $\omega_c$ the frequency carrier (in rads/s). The $L_s$ is a factor representing system losses which are inherent to hardware, and not related to propagation issues (assuming that there are no losses in the system, we will consider a value of $L_s = 1$).

By definition, the antenna's gain is related to the antenna's effective area or "aperture" by,

$$G = \frac{4\pi A_e}{\lambda^2},\tag{2.3}$$

where the aperture $A_e$ is related to the dimensions of the antenna. However, in wireless systems isotropic antennas are used in order to have reference antenna gains. An isotropic radiator consists of an ideal antenna which transmits energy uniformly in all directions, having unit gain ($G = 1$).

It can be seen from equation (2.1) that the received power falls off with the square of the distance $d$, which can be quantified as a decay with distance at a rate of 20 dB/decade [Rap01].

In fact, the signal attenuation of an electromagnetic wave represented by the *path-loss*, is measured in dB, and it gives the difference between the effective isotropic radiated power (EIRP) and the received power, and consists of a theoretical measurement of the maximum radiated power available from a transmitter in the direction of maximum antenna gain, as compared to an isotropic radiator. The *path-loss* for the free space model is given by

$$PL = 10 \log \frac{P_t}{P_r} = -10 \log \left[ \frac{G_t G_r \lambda^2}{(4\pi)^2 d^2} \right], \tag{2.4}$$

It can be seen that in free-space, we have

$$PL \propto d^2. \tag{2.5}$$

However, in practical scenarios in which the transmitted signal may be reflected, the power signal decays faster with distance. Several propagation models show that the average received signal power decreases in a logarithmical form with distance. And it is defined that the average large-scale *path-loss* (over an infinity of different points) for a distance $d$ between the transmitter and receiver can be given as a function of $d$ with resort to *path-loss* factor $n$, which is the rate at which the *path-loss* increases with distance. Therefore, a simple model for *path-loss* given by [Rap01], may be

$$\overline{PL} = \overline{PL}(d_0) + 10n \log \left( \frac{d}{d_0} \right), [dB] \tag{2.6}$$

where $\overline{PL}(d_0)$ is the mean *path-loss* in dB at distance $d_0$ (the bar in the equation refers to the joint average of all possible loss values). It is also important to point out that since equation (2.1) is not defined for $d = 0$, then the term $d_0$ is used as a known received power reference point [Rap01]. Hence, the received power $P_r(d)$, at a given spatial separation $d > d_0$, may be related to $P_r(d_0)$. This reference point can be obtained analytically with resort to equation (2.1), or experimentally by measuring the received power in several locations sited in a radial distance $d_0$ from the transmitter, and performing the average. Typically, depending on the size of the covered area, $d_0$ is assumed to be 1 *km* for large cells and 100 *m* for microcells. The linear regression for a minimum mean-squared estimate (MMSE) that fits $\overline{PL}$ versus $d$ on a log-scale produces a direct line with constant decay of 10 dB/decade (which in free-space, with $n = 2$, results in the 20 dB/decade slope mentioned above).

## 2.1.2   Shadowing

Another type of large-scale fading is called *shadowing*, and it is caused by obstacles (e.g., clusters of buildings, mountains, etc.) between the transmitter and receiver. As a result a portion of the transmitted signal is lost due to reflection, scattering, diffraction, and even absorption.

It is important to note that equation (2.6), which defines the *path-loss* versus distance $d$, represents an average, and therefore it might not be appropriate to describe the attenuation of a particular path. Hence, as a result of *shadowing*, the received power in two different locations at the same distance $d$ from the transmitter may have very different values of *path-loss* than the ones predicted by equation (2.6). Since the environment of different locations with the same distance $d$ may be different, it is therefore necessary to introduce variations about the mean loss defined in equation (2.6).

Several measurements in real scenarios have shown that the *path-loss PL* at a given distance $d$ is a random variable characterized by a log-normal distribution about the mean value $\overline{PL}$ [Skl97]. Therefore the measured *path-loss PL* (in dB) varies around the distance-dependent mean loss (given by equation (2.6)), and therefore $LP$ can be written in terms of $\overline{PL}$ plus a random variable (R.V.), $X_\sigma$, by [Rap01]

$$\overline{PL} = \overline{PL}(d) + X_\sigma = \overline{PL}(d_0) + 10n \log(\frac{d}{d_0}) + X_\sigma, [dB] \qquad (2.7)$$

where, $X_\sigma$ is a normal (or Gaussian) distributed random variable (RV) with zero mean and standard deviation $\sigma$, and this RV represents the effect of shadowing. This distribution is suitable to define the random effects associated with the *log-normal shadowing* phenomenon, and it considers that several different measure points at a given distance $d$ have a Gaussian distribution around the mean loss defined in equation (2.6) [Rap01]. In sum, in order to statistically define the path-loss caused by large-scale fading for a given distance $d$, a set of values has to be defined: *path-loss* exponent, reference point $d_0$, standard deviation $\sigma$ of the RV $X_\sigma$.

## 2.2   Small-Scale Fading

A very important type of fading normally considered in wireless communication systems is related to rapid changes in the signal's amplitude and phase that occur over very short variations in time or in the spatial area between the receiver and the transmitter (in fact, drastic changes in signal strength may be noticed even in half a meter shift). The propagation model that describes this type of fading is called small-scale fading and can be expressed by two factors [Rap01]:

■ Delay spread $T_m$, due to the multipath propagation. It is related to frequency selectivity which in the time domain translates in time dispersion of the signal;

■ Doppler spread $B_D$, due to relative motion between the transmit and/or receive antenna. It is related to time selectivity which in the frequency domain translates in frequency dispersion of the signal frequency components.

Moreover, in mobile transmission the velocity also plays an importance role in the type of fading experienced by the signal.

Different transmitted signals are subjected to different effects of fading. In fact, the type of fading "sensed" by the transmitted signal is defined by a relation between the properties of the signal and the characteristics of the channel. Depending on these characteristics, a set of different effects of small-scale fading can be experienced. As we will see next, while multipath delay spread leads to time dispersion and frequency selective fading, Doppler spread leads to frequency dispersion and time selective fading. The two propagation mechanisms are independent of each other. The propagation models characterizing these rapid fluctuations of the received signal amplitude over very short time durations are called small-scale fading models.

## 2.2.1 The Multipath Channel

The main difference between a wired and wireless communication system lies in the propagation environment. In a wired communication system there is only a single path propagation between the transmitter and the receiver. On the other hand, wireless communication can be affected by distinct natural phenomena like interference, noise, and other factors that represent serious impairments. Since most mobile communication systems are used within urban environments, a major constraint is related to the fact that the mobile antenna is well below the height of the nearby structures (such as cars, buildings, etc.), and as a consequence, the radio channel is influenced by those structures. In fact, since within this type of scenario the line-of-sight component does not usually exist, communication is only possible due to the influence of propagation mechanisms (reflection, diffraction, and scattering of the multipath waves). The wireless communication system is characterized by a multipath propagation environment, a phenomenon in which the incoming multipath components arrive at the receiving antenna by different propagation paths, giving rise to different propagation time delays and leading the signal to fade. Multipath fading may be caused by a set of effects which significantly affect signals' propagation in wireless transmission, such as reflection, diffraction, and scattering.

**Reflection** occurs when an electromagnetic wave encounters a surface that is large relative to the wavelength of the propagation wave (e.g., walls of a building, hills, and other large plain surfaces), and it is illustrated in Fig. 2.1.

**Diffraction** occurs when the path between the transmitter and receiver is obstructed by an object with large dimensions when compared to the wavelength of the propagation wave, being diffracted on the edges of such objects (e.g., cars, houses, mountains). The wave tends to travel around the object allowing the signal to be received, even if the receiver is shadowed by the large object. It is illustrated in Fig. 2.2.

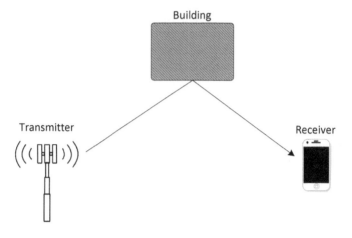

**Figure 2.1: Reflection effect.**

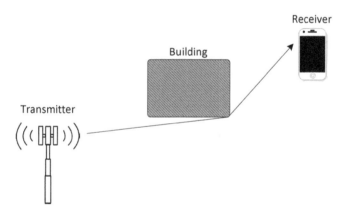

**Figure 2.2: Diffraction effect.**

**Scattering** occurs when an incoming signal hits an object whose size is in the order of the wavelength of the signal or less. Scattering waves are usually produced by rough surfaces or small objects (e.g., road signs, lamp posts, foliage, etc.). The radio signal undergoes scattering on a local scale, and it is typically characterized by a large number of reflections in small objects in the mobile's vicinity. It is illustrated in Fig. 2.3.

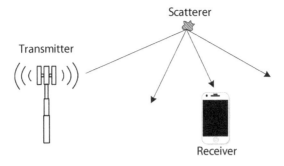

**Figure 2.3: Scattering effect.**

In a multipath propagation environment, several copies of the transmitted signal arriving from different paths, and having different delays, combine at the receiver with different attenuations. Furthermore, depending on the delay, each incoming signal will have a different phase factor. Depending on their relative phases, these multipath components will add up constructively or destructively, causing fluctuations in the overall received signal's amplitude. And depending on the addition of the signal copies across the received path, the receiver will see a single version of the transmitted signal with a corresponding gain (attenuation) and phase. While constructive interference affects the overall signal positively since it increases the amplitude of the overall signal, destructive interference is caused by mutual cancelation of different multipath components leading to a decrease of the signal level.

If the multipath fading channel has very long path lengths, then copies of the original signal may arrive at the receiver after one symbol duration, which will interfere with the detection of the posterior symbol, resulting in inter-symbol interference (ISI), and thus inducing distortion which causes significant degradation of the performance of the transmission. In this case, the multipath components will no longer be separable in time. Still they can be separated in frequency, and therefore the inter-symbol interference can be compensated with resort to frequency domain equalization. This will be seen later in the subsequent chapters.

Assume that the transmitter transmits a very short pulse over a multipath channel.

Two important parameters have to be taken into account: the symbol's time duration, $T_S$, and the delay spread $T_m$ (also known as maximum excess delay time). The delay spread is a fundamental parameter in the characterization of the multipath fading, since it defines the time elapsed between the first received component and the last (in order to define the relevant components a threshold is usually chosen at 20 dB below the strongest multipath component). On other words, it represents the length of the impulse response of the channel. Hence, considering a certain symbol with period $T_S$ is transmitted, the symbol will be spread out by the channel, and at the receiver side its length will be the $T_S$ added to the delay spread $T_m$. Depending on the relation between $T_S$ and $T_m$, the degradation can be classified in two types: flat fading or frequency selective fading. Basically, if the delay spread is much smaller than the symbol period then the channel exhibits flat fading and the ISI can be neglected. On the other hand, if the delay spread is equal to or greater than the symbol period, the channel introduces ISI that must be compensated.

**Frequency-Selective Fading**

The multipath channel introduces time spread in the transmitted signal, since due to multipath reflections, the channel impulse response will appear as a series of pulses, as illustrated in Fig. 2.4. The multipath components may sum constructively or destructively, and the receiver sees an overall single copy of the transmitted signal, characterized by a given gain (i.e., attenuation) and phase. If the channel impulse response has a delay spread $T_m$ greater than the symbol period $T_S$ of the transmitter signal, (i.e., $T_m > T_S$), then the dispersion of the transmitted symbols within the channel will lead to ISI causing distortion on the received signal.

In the frequency domain, the spectrum of the received signal shows that the bandwidth of the transmitted signal is greater than the coherence bandwidth of the channel, and in these conditions the channel induces frequency-selective fading over the bandwidth. A channel parameter called coherence bandwidth, $B_C$, is used to characterize the fading type. It consists of a statistical measure of the frequency bandwidth in which the channel characteristics remain similar (i.e., "flat"). Essentially, signals with frequencies separated by less than $B_C$ will experience very similar gains. A signal undergoes flat fading if in the frequency domain $B_S \ll B_C$, and in the time domain $T_S \gg T_m$, as illustrated in Fig. 2.5(a).

On the other hand, if the bandwidth of the transmitted symbol is greater than the channel coherence bandwidth, $B_S > B_C$, and in the time domain $T_m > T_S$, then different frequency components of the signal experience different fading. In such conditions, the spectrum of the received signal with different frequency components will have some components with greater gains than others. Thus, frequency-selective fading causes distortion of the transmitter signal since the signal's spectral components are not all affected in the same way by the channel. In fact, as can be seen in Fig. 2.5(b) when the signal's bandwidth is greater than the coherence bandwidth of the channel, the spectral components placed within the coherence bandwidth will be affected in a different way when compared to the components that are not covered by it. It is important to note that the coherence bandwidth $B_C$, and the delay spread $T_m$, can be related by $B_C = \frac{1}{T_m}$. Thus, the coherence bandwidth and the

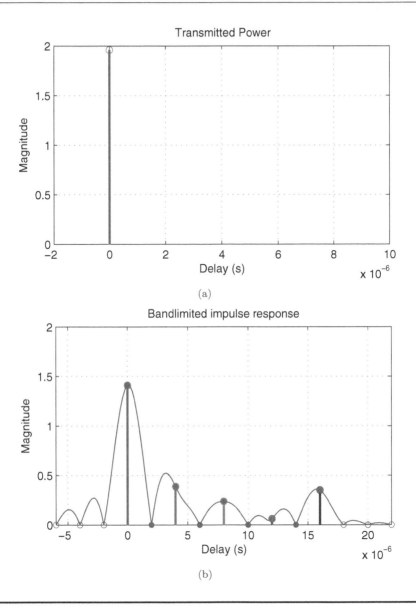

**Figure 2.4: Multipath power delay profile: Power transmitted (a); channel impulse response (b).**

delay spread are inversely related: the larger the delay spread, the less the coherence bandwidth and the channel is said to be more frequency selective. Hence, multipath propagation, leads to frequency selective fading.

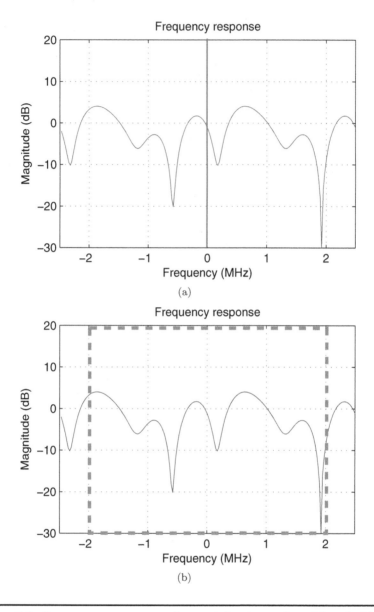

**Figure 2.5: Frequency response of a certain channel and bandwidth of signal (dotted line): Narrowband signal (a); wideband signal (b).**

## *Modeling the Multipath Channel*

We wish to model the wireless propagation channel as the system illustrated in Fig. 2.6.

**Figure 2.6: Transmission of a symbol $s(t)$ through wireless channel $h(t)$.**

Assume that the system has an impulse response $h(t)$, and that a transmitter sends a signal $s(t)$, (which in Fig. 2.6 is presented at the system's input). Assume that $s(t)$ propagates through a wireless channel characterized by a response $h(t)$, with the output of the signal corresponding to the received signal at the receiver. In a linear time invariant systems we have: if the signal $s(t)$ is passed through $h(t)$, the output $y(t)$ will be the convolution between $s(t)$ and $h(t)$. From the theory of linear systems, we know that an attenuation is simply a scaling of the signal, which corresponds to multiplying the signal by a scaling constant denoted by the attenuation factor (or gain), and is denoted by $\alpha_l$. On the other hand, a delay simply corresponds to an impulse function $\delta(t - \tau_l)$, where $\tau_l$ is the respective delay.

Let us look at the mobile communication system illustrated in Fig. 2.7 Regarding the $0^{th}$ path, the signal $s(t)$ is attenuated by $\alpha_0$ and delayed by $\tau_0$, and this can be represented as a system with impulse response $\alpha_0\delta(t - \tau_0)$. In the same way, the $1^{st}$ path can be described by an attenuation $\alpha_1$ and a delay $\tau_1$, with that corresponding path being represented by $\alpha_1\delta(t - \tau_1)$. Similarly for the $2^{nd}$ path we have $\alpha_2\delta(t - \tau_2)$, and so on and so forth, until the $L^{th}$ path. Having characterized the $L$ paths (the $0^{th}$ path corresponding to the *LOS* component plus $L - 1$ scattered components), we can model the wireless

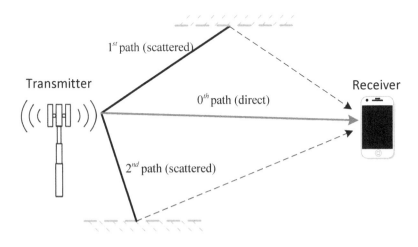

**Figure 2.7: Mobile communication system.**

channel as a combination of all paths as

$$h(t) = \alpha_0\delta(t-\tau_0)+\alpha_0\delta(t-\tau_0)+\ldots+\alpha_{L-1}\delta(t-\tau_{L-1}) = \sum_{l=0}^{L-1}\alpha_l\delta(t-\tau_l) \quad (2.8)$$

and it becomes clear that the wireless channel impulse response $h(t)$ can be given by the sum of the impulse responses of the $L$ paths. Thus, the wireless channel can be represented as a sum of multipath components each one characterized by a given attenuation and delay, as in Fig. 2.4. The multipath fading channel is therefore modeled as a linear finite impulse-response (FIR) filter. The channel's filtering behavior is caused by the sum of amplitudes and delays of the multipath components at the same instant in time. The channel will behave as a filter, whose frequency response exhibits frequency selectivity. In the frequency domain, this refers to the case in which the bandwidth of the transmitted symbol is greater than the channel coherence bandwidth. Then the signal will suffer from frequency selective fading. Different components of the transmitted signal will suffer from different attenuations and therefore will have different gains. Thus, frequency-selective fading causes the distortion of the transmitter signal since the signal's spectral components are not all affected in the same way by the channel. The coherence bandwidth and the delay spread are inversely related: the larger the delay spread, the less the coherence bandwidth and the channel is said to be more frequency selective. Hence, multipath propagation leads to the time dispersion of the transmitted symbol within the channel, leads to frequency-selective fading. In fact, the demand for higher data rates will face an increase of the frequency selectivity of the wireless channel.

## Modeling the Transmitted Signal

Having modeled the wireless propagation channel, it is important to characterize the transmitted signal $s(t)$ through the wireless channel.

Typically the signal to be transmitted is band-limited, i.e., defined in $[-B, B]$ Hz and zero elsewhere. Such signals which primarily occupy a range of frequencies centered around 0 Hz are called low-pass (or baseband) signals.

The majority of the wireless communication systems use modulation techniques in which the information bearing baseband signals are upconverted using a sinusoidal carrier of frequency $f_c$ before transmission, in order to move the signals away from the DC component and center them in an appropriate frequency carrier. The resulting transmitted signal $s(t)$ will have a spectrum $S(f)$ which is zero outside of the range $f_c - B < |f| < f_c + B$, where $f_c >> B$ (i.e., the bandwidth $B$ of the spectrum $S(f)$ is much smaller than the carrier frequency $f_c$). The term passband is often applied to these signals. Considering a communication system in which the involved signals can be measured by the received antenna (i.e., typically voltage signals), the transmitted signals and received signals are typically referred to as real passband signals.

However, the process of modulation which aims to form a signal suitable for transmission, requires an operation to translate the baseband message signal to a passband signal, located in much higher frequencies when compared with the baseband frequency. Since the information is contained in the modulated signal and not in the carrier frequency used for the transmission, then a model to define the signal $s(t)$ independently of the carrier frequency $f_c$, must be derived. The passband transmitted and received signals are therefore converted to the corresponding equivalent baseband signal, which is processed by the receiver in order to recover the information. Hence, a very simple representation was developed to achieve this, and is called *complex baseband* or *low-pass equivalent* representation of the communication (passband) signal, and is detailed in Appendix B. Therefore, the signal $s(t)$ consists of a passband signal transmitted at the carrier frequency, and is written as

$$s(t) = Re\left\{ s_b(t)e^{j2\pi f_c t} \right\},$$

where $s_b(t)$ corresponds to the complex baseband representation of $s(t)$. The real and imaginary parts of the complex quantity $s_b(t)$ carry information about the signal's inphase and quadrature components (the components that are modulating the terms $\cos(2\pi f_c t)$ and $\sin(2\pi f_c t)$, respectively).

## Modeling the Received Signal

Let us now derive the received signal at the receiver, after passing through the wireless channel. The received signal $y(t)$ consists of a convolution of the transmitted signal $s(t)$ with the wireless channel $h(t)$. In order to better understand this process, let us derive component by component. Passing the signal $s(t)$ through the $0^{th}$ component (LOS), given by $\alpha_0\delta(t - \tau_0)$, then the signal will be attenuated by $\alpha_0$ and delayed by $\tau_0$. Hence, the received signal corresponding to this specific path is simply,

$$y_0(t) = Re\left\{ \alpha_0 s_b(t - \tau_0)e^{j2\pi f_c(t-\tau_0)} \right\}, \tag{2.9}$$

In the same way, the signal corresponding to the $1^{st}$ component is given by

$$y_1(t) = Re\left\{ \alpha_1 s_b(t - \tau_1)e^{j2\pi f_c(t-\tau_1)} \right\}. \tag{2.10}$$

The same procedure is repeated for the rest of the components, with the signal corresponding to the $L - 1$ paths being given by

$$y_{(L-1)}(t) = Re\left\{ \alpha_{(L-1)} s_b(t - \tau_{(L-1)})e^{j2\pi f_c(t-\tau_{(L-1)})} \right\}. \tag{2.11}$$

The overall received signal can be represented as the sum of all signal contributions, i.e.,

$$y(t) = Re\left\{ \sum_{l=0}^{L-1} \alpha_l s_b(t - \tau_l)e^{j2\pi f_c(t-\tau_l)} \right\}. \tag{2.12}$$

Clearly, the carrier term given by $e^{j2\pi f_c t}$ is common to all terms. By isolating $e^{j2\pi f_c t}$ then (2.12) can be rewritten as

$$y(t) = Re\left\{\left(\sum_{l=0}^{L-1} \alpha_l s_b(t - \tau_l)e^{-j2\pi\tau_l}\right)e^{j2\pi f_c t}\right\}. \qquad (2.13)$$

In (2.13), the term inner brackets consists of a complex baseband received signal, hence (2.13) can be rewritten as

$$y(t) = Re\left\{y_b(t)e^{j2\pi f_c t}\right\} \qquad (2.14)$$

where the equivalent complex baseband representation of $y(t)$ is

$$y_b(t) = \sum_{l=0}^{L-1} \alpha_l s_b(t - \tau_l)e^{-j2\pi\tau_l} \qquad (2.15)$$

and the equivalent lowpass representation of the channel is

$$h_b(t) = \sum_{l=0}^{L-1} \alpha_l e^{-j2\pi\tau_l}\delta(t - \tau_l). \qquad (2.16)$$

The complex baseband received signal at the receiver, given by $y_b(t)$, consists of the sum of the $L$ received multipath components, each one having a corresponding attenuation $\alpha_l$, and delay $\tau_l$.

Let us assume that the baseband signal of different values of $\tau_l$ is approximately $s_b(t)$; therefore, (2.13) can be simplified with resort to the narrowband assumption, since all the terms $s_b(t - \tau_l)$ are approximately equal to $s_b(t)$, this is,

$$y_b(t) = s_b(t)\sum_{l=0}^{L-1} \alpha_l e^{-j2\pi\tau_l}. \qquad (2.17)$$

and we reach a point at which an analytical model of the wireless transmission system can be defined as

$$y_b(t) = h_b(\tau)s_b(t). \qquad (2.18)$$

Let us focus on the equivalent lowpass representation of the channel, $h_b(\tau)$, where $\tau$ corresponds to a given delay. A fundamental factor can be observed from the above expression. We will call it phase factor, and it denotes the term given by $e^{-j2\pi\tau_l}$. It has been explained before that since the different multipath components travel through different distances, they are received with different delays. The delay induces a phase at the signal received relative to the $l$th multipath component, and it is clear that the phase factor $e^{-j2\pi\tau_l}$ arises out of the delay $\tau_l$. As a result of the different delays, the multipath components sum up with different phases at the receiver.

Depending on the different attenuations and delays, the summation of the different components can produce destructive or constructive interference. Since it is impossible to know the exact values of the attenuation and delay for all of the multipath components in real time wireless transmissions, a statistical approach can be taken in order to understand the properties of the complex fading coefficient. Hence, instead of trying to characterize each component separately, it is possible to describe the properties of the equivalent lowpass representation of the channel as a whole, with resort to the theory of random processes, statistics and probability. The statistical characteristics of the channel exhibiting small-scale fading can be modeled by several probability distribution functions. Notwithstanding the existence of a large number of scatterers within the channel (contributing to the received signal), and assuming that the different scatterers are independent, the central limit theorem (CLT) can be used to approximate the components as independent Gaussian R.V.'s, therefore allowing us to model the channel impulse response as a Gaussian process.

Let us statistically analyze the equivalent lowpass representation of the channel, in order to draw some conclusions about its random behavior. We can apply a small modification to (2.18) in order to write it as a sum of the real part and imaginary part, i.e.,

$$y_b(t) = h_b(\tau)s_b(t) = s_b(t) \sum_{l=0}^{L-1} \alpha_l e^{-j2\pi f_c \tau_l}$$

$$= s_b(t) \sum_{l=0}^{L-1} \alpha_l \cos(2\pi f_c \tau_l) - \alpha_l \sin(2\pi f_c \tau_l),$$

(2.19)

where the real part and imaginary parts of this complex-valued quantity are given by (2.20) and (2.21), respectively,

$$X = \left( \sum_{l=0}^{L-1} \alpha_l \cos(2\pi f_c \tau_l) \right), \tag{2.20}$$

$$Y = \left( \sum_{l=0}^{L-1} -\alpha_l \sin(2\pi f_c \tau_l) \right). \tag{2.21}$$

Here $X$ and $Y$ are both random numbers depending on the random quantities given by $\alpha_l$ and $\tau_l$. The randomness of these components is due to the fact that each component is arising from the multipath environment. The wireless channel can therefore be analyzed with resort to statistical propagation models, where the channel parameters are modeled as stochastic variables. Hence, $X$ and $Y$ are derived as the sum of a large number of random components, and in these conditions we can assume that $X$ and $Y$ are both Gaussian random variables. Hence, $h_b(\tau)$ can be rewritten as

$$h_b(\tau) = X + jY, \tag{2.22}$$

Considering that $X$ and $Y$ are Normal-distributed, then

$$X \sim N(0, 1/2) \tag{2.23}$$

$$Y \sim N(0, 1/2). \tag{2.24}$$

That is, $X$ and $Y$ can be described as Gaussian random variables of mean zero and variance $\frac{1}{2}$. Assuming that the process has zero mean, the envelope of the received signal can be statistically described by a Rayleigh probability distribution, with the phase uniformly distributed in $(0, 2\pi)$. Hence, assuming that $X$ and $Y$ are independent R.V.'s the joint distribution of $XY$ can be expressed by the product of the individual distributions of $X$ and $Y$, which are given by

$$
\begin{aligned}
f_X(x) &= \frac{1}{\sqrt{2\pi}\sigma} e^{-\frac{(x-\mu)^2}{2\sigma^2}} \\
&= \frac{1}{\sqrt{\pi}} e^{-x^2},
\end{aligned}
\tag{2.25}
$$

and

$$
\begin{aligned}
f_Y(y) &= \frac{1}{\sqrt{2\pi}\sigma} e^{-\frac{(y-\mu)^2}{2\sigma^2}} \\
&= \frac{1}{\sqrt{\pi}} e^{-y^2}.
\end{aligned}
\tag{2.26}
$$

with the joint distribution given by

$$
\begin{aligned}
f_{X,Y}(x,y) &= f_X(x) \cdot f_Y(y) \\
&= \frac{1}{\sqrt{\pi}} e^{-(x^2+y^2)}
\end{aligned}
\tag{2.27}
$$

which allows to obtain the joint distribution of the components of $h_b(\tau)$.

### 2.2.2   Time-Varying Channel

Besides multipath propagation, time variations within the channel may also arise due to oscillator drifts, as well as due to mobility between transmitter and receiver. Oscillator drifts consist of frequency errors relative to the frequency mismatch between the local oscillator at the transmitter and the local oscillator at the receiver and can be caused by phase noise or residual CFO. These channel variations lead to simple phase variations that are relatively easy to compensate at the receiver [SF08, DAPN10].

#### Time Variation Due to Carrier Frequency Offset

The carrier frequency offset results from a mismatch between the local oscillator at the transmitter and the local oscillator at the receiver and can lead to performance degradation. In order to better understand how the CFO affects the coherent detection of the transmitted signal, let us illustrate a scenario in which a transmitter sends a signal $s(t)$ that passes through the channel and is recovered by the receiver. Assume the existence of a mismatch between the local oscillator at the transmitter and the local oscillator at the receiver, in which the transmitter sends $s(t)$ over a carrier $f_c + \Delta_f$ when it should use $f_c$. On the receiver side, the local oscillator is tuned to the reference carrier frequency $f_c$. How will this affect the received signal? Consider that the transmitted signal is given by

$$s(t) = Re\left\{s_b(t)e^{j2\pi(f_c+\Delta_f)t}\right\},$$

with $s_b(t)$ denoting the baseband representation of $s(t)$. However, the receiver is not aware of the frequency offset at the transmitter side. It interprets the baseband representation of $s(t)$ as $s^\Delta(t) = s_b(t)e^{j2\pi\Delta_f t}$ (when it is not), and interprets the term as the carrier frequency $e^{j2\pi(f_c)}$. The received signal will be written as

$$
\begin{aligned}
y(t) &= Re\left\{\int_{-\infty}^{\infty} h_b(t,\tau)s_b(t-\tau)d\tau e^{j2\pi f_c t}e^{j2\pi\Delta_f t}\right\} \\
&= e^{j2\pi\Delta_f t}Re\left\{\int_{-\infty}^{\infty} h_b(t,\tau)s_b(t-\tau)d\tau e^{j2\pi f_c t}\right\},
\end{aligned}
\tag{2.28}
$$

and the equivalent baseband is given by

$$y_b(t) = e^{j2\pi\Delta_f t}\int_{-\infty}^{\infty} h_b(t,\tau)s_b(t-\tau)d\tau, \tag{2.29}$$

which in the discrete time domain is given by

$$y_n = e^{j2\pi\theta_n}\sum_{l=0}^{L-1} h_{n,l}s_{n-l}, \tag{2.30}$$

where the CFO is given by $\theta_n = \Delta_f T$. Looking to the received signal in the discrete time it becomes very clear that after demodulation at the receiver, this frequency mismatch results in a time varying phase which is multiplied by the received signal. It is also very important to note that the CFO is common to all propagation paths, therefore all components are affected by the same frequency shift. Therefore, the spectrum of the received signal is shifted in frequency (and not broadened, in opposition to the Doppler spread in which each wave experiences a different frequency shift depending on the angle of incidence, as will be seen next). Hence, the equivalent channel has a phase rotation that changes with time, and this is the reason the channel affected by CFO is said to vary in time.

## Time Variation Due to Movement

We have seen before that the wireless channel can be described as a function of time (and space), with the equivalent received signal resulting from the combination of the different replicas of the original signal arriving at the receiving antenna by different propagation paths and delays. These multipath components will suffer from interference in a constructive or destructive way, depending on their relative phases. As a consequence, this effect will cause fluctuations in the received signal. Now, considering that either the transmitter or the receiver is moving, this propagation phenomenon will be time varying. When there is relative motion between the mobile and the fixed base station, the multipath components experience an apparent shift in frequency, called Doppler shift (dependent on the mobile speed, carrier frequency, and the angle that its propagation vector makes with the direction of motion). Small-scale fading based on Doppler spread can be classified in fast fading or slow fading channel, depending on how rapidly the transmitted baseband signal changes as compared to the rate of variation of the channel.

The channel is said to exhibit slow fading if the channel impulse response changes at a rate very much slower than the transmitted symbol time, i.e., $T_C \gg T_S$, where $T_C$ stands for the coherence time of the channel. When the channel impulse response changes rapidly within the symbol duration then $T_C$ will be smaller than the time duration of a transmission symbol, $T_S$, such as $T_C < T_S$ [Rap01]. In these conditions, fast fading will arise. If the channel keeps changing during the time in which a symbol is propagating it will lead to the distortion of the baseband pulse shape, resulting in a loss of SNR which can cause a high error rate, as well as synchronization problems. This distortion occurs due to the fact that the received signal's components are not all highly correlated throughout time [Rap01]. Looking at the frequency domain, frequency dispersion arises due to Doppler spreading. The signal distortion due to fast fading increases with increasing Doppler spread with regard to the bandwidth of the transmitted signal [Rap01]. Hence, a signal undergoes fast fading if in the time domain $T_C < T_S$, and in the frequency domain $B_S < B_D$. Frequency dispersion is considered a major impairment in mobile communications.

One of the most important challenges for mobile communications systems is to overcome the severe effects of the mobile radio channel. For instance, a typical transmission between a moving vehicle and a fixed based station within an urban environment is subjected to extreme fluctuations in both amplitude and frequency, where fades up to 40 dB below the mean level can frequently occur, and in very short time intervals (about every half wavelength of the frequency carrier). As a consequence of motion, the receiving antenna experiences strong random signal fluctuations due to the random distribution of the propagation channel in space. Consequently, this effect will distort the received signal. The constructive or destructive interference at the receiver is different for each position in space. If the receiver is moving, the channel varies with location and time, and therefore at each position in time, the receiver will feel a different signal interference. This leads to a type of fading in which the received amplitude and phase both vary in time.

The relative motion between the transmitter and the receiver also results in Doppler frequency shift, leading to channel variations which are not easy to compensate. The Doppler effect has a strong negative impact on the performance of mobile radio communication systems since it causes a different frequency shift for each incident plane wave. Fig. 2.8 illustrates the transmission through a channel characterized by multipath propagation, between a mobile transmitter traveling with speed $v$, and a fixed receiver. Due to the

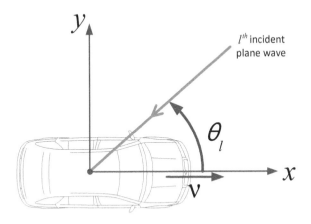

**Figure 2.8: Example of a $l^{th}$ incident wave affected by the Doppler effect.**

relative movement between the transmitter and receiver, the frequency of the received signal suffers from a Doppler frequency shift, which is proportional to the speed of the transmitter and to the spatial angle between the direction of the movement and the direction of departure and/or arrival of the component. Each multipath component experiences its own Doppler shift. The Doppler shift is different for each propagation path since it depends on the angle of incidence in relation to the direction of motion. Hence, paths associated to a receiver moving away from the transmitter experience a decrease in frequency (negative Doppler), while paths traveling into experience an apparent increase in frequency (positive frequency). Multipath components arriving in the intermediate angle will suffer the corresponding shift.

$$f_l = f_c + f_D^{(l)}, \qquad (2.31)$$

with the Doppler shift associated with the $l^{th}$ multipath component denoted by

$$f_D^{(l)} = \frac{v}{c} f_c \cos(\theta_l) = f_D \cos(\theta_l), \qquad (2.32)$$

where $f_D = v f_c / c$ represents the maximum Doppler shift (which increases linearly with the carrier frequency $f_c$ and the speed of the mobile $v$), and $\theta_l$ is the angle between the direction of the motion and the directions of arrival of the $l^{th}$ multipath component. The maximum Doppler shift occurs when $\theta_l = 0$, while the minimum occurs when $\theta_l = \pm\pi$. On the other hand, the

Doppler shift $f_D^{(l)} = 0$ if $\theta_l = \pi/2$ or $\theta_l = 3\pi/2$. In the frequency domain the spectrum of the transmitted signal experiences a frequency distension, a phenomenon which is known as frequency dispersion. The extension of the frequency dispersion depends on the amplitudes of the received waves and the maximum Doppler shift.

Therefore, in mobility scenarios, the rate of variation of the channel response in time is characterized by the Doppler spread. Significant variations of the channel response within the signal duration lead to **time-selective fading**, and this represents a major impairment in wireless communication systems. In fact, these time variations are unpredictable which means that the time-varying nature of multipath channel must be characterized statistically.

In order to describe the time-varying channel impulse response, let us consider the transmitted signal $s(t)$, given by

$$s(t) = Re\left\{s_b(t)e^{j(2\pi f_c t+\phi_0)}\right\} \tag{2.33}$$

where $\phi_0$ denotes the phase offset of the carrier.

The transmitted signal propagates over the local scatterers via several paths and arrives at the receiver antenna coming from various directions. Figure 2.9 illustrates a multipath propagation scenario between a radio transmit-

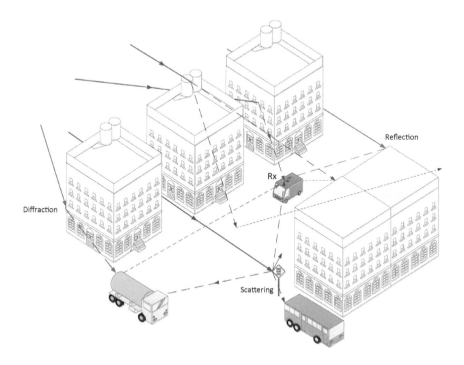

**Figure 2.9: A mobile receiver within a multipath propagation scenario.**

ter (which we assume to be stationary) and a moving receiver, in a presence of multiple reflectors.

A different attenuation and a phase shift will be caused by each scatterer, and since each individual path is characterized by a propagation delay due to the random nature of the channel, both the attenuation and delay will be time variant. A signal $s(t)$ propagating through the time-varying channel will be received as

$$
\begin{aligned}
y(t) &= Re\left\{\sum_{l=0}^{L-1} \alpha_l(t)s_b(t - \tau_l(t))e^{j2\pi f_l(t-\tau_l(t))}\right\} \\
&= Re\left\{\sum_{l=0}^{L-1} \alpha_l(t)s_b(t - \tau_l(t))e^{j2\pi\left(f_c + f_D^{(l)}\cos(\theta_l)\right)(t-\tau_l(t))}\right\} \\
&= Re\left\{\sum_{l=0}^{L-1} \alpha_l(t)s_b(t - \tau_l(t))e^{j2\pi\left(f_c t - f_c\tau_l(t) + f_D^{(l)}\cos(\theta_l)t - f_D^{(l)}\cos(\theta_l)\tau(t)\right)}\right\} \\
&= Re\left\{\sum_{l=0}^{L-1} \alpha_l(t)s_b(t - \tau_l(t))e^{j2\pi\left(f_c(t-\tau_l(t)) + \phi_{Dop,l} - f_D^{(l)}\cos(\theta_l)\tau(t)\right)}\right\} \\
&= Re\left\{\sum_{l=0}^{L-1} \alpha_l(t)s_b(t - \tau_l(t))e^{j2\pi f_c t - j2\pi\left(f_c\tau_l(t) - \phi_{Dop,l} + f_D^{(l)}\cos(\theta_l)\tau(t)\right)}\right\} \\
&= Re\left\{\sum_{l=0}^{L-1} \alpha_l(t)s_b(t - \tau_l(t))e^{j2\pi f_c t - \phi_l(t)}\right\},
\end{aligned}
$$

$$(2.34)$$

where $\phi_{Dop,l}$ stands for the Doppler phase shift, and a simplification of the phase factor given by $\phi_l(t) = j2\pi\left(f_c\tau_l(t) - \phi_{Dop,l} + f_D^{(l)}\cos(\theta_l)\tau(t)\right)$. From (2.34) can be taken the equivalent lowpass representation of $y(t)$, given by

$$
y_b(t) = \sum_{l=0}^{L-1} \alpha_l(t)s_b(t - \tau_l(t))e^{-j\phi_l(t)} \tag{2.35}
$$

The result of (2.35) clearly highlights the propagation effects over a multipath channel. Considering a signal transmitted through a time-varying multipath channel, the equivalent lowpass received signal seems like a sum of attenuated and delayed versions of the original signal. The attenuations are complex-valued and time-variant. This multipath characteristic of the channel causes the transmitted signal to "extend" in time, and as a consequence the received signal will have a greater duration than the transmitted signal, a phenomenon known as *time dispersion*. This representation can be interpreted as a transversal filter of order $L$ with time-varying tap gains. Fig. 2.10 illustrates the tapped delay line model of a doubly-selective channel in the equivalent complex baseband. Modeling this type of fading channels can

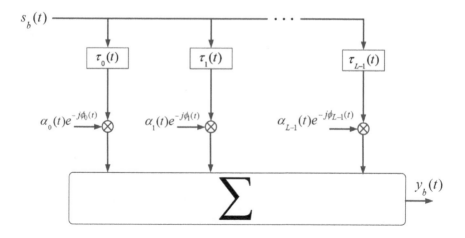

**Figure 2.10: Tapped delay line model of a doubly-selective channel in the equivalent complex baseband.**

represent a difficult task since each one of the different multipath components must be modeled, and the mobile radio channel has to be modeled as a linear filter having a time varying impulse response $h(t, \tau)$.

It has been shown that the output signal $y(t)$ is the result of the convolution of the baseband input signal, $s_b(t)$ with the time-varying channel impulse response. From (2.35) an input/output relationship comes in evidence, which allows us to think that the time-variant impulse response of the channel can be derived from it. Hence, by doing a small simplification in (2.35) we get

$$
\begin{aligned}
y(t) &= \mathrm{Re}\left\{ \left( \sum_{l=0}^{L-1} \alpha_l(t) s_b(t - \tau_l(t)) e^{-j\phi_l(t)} \right) e^{j2\pi f_c t} \right\} \\
&= \mathrm{Re}\left\{ \left( \int_{-\infty}^{\infty} h(t, \tau) s_b(t) d\tau \right) e^{j2\pi f_c t} \right\},
\end{aligned}
\tag{2.36}
$$

where the equivalent lowpass representation of the time-varying channel can be expressed as

$$
h_b(t, \tau) = \sum_{l=0}^{L-1} \alpha_l(t) \delta(t - \tau_l(t)) e^{-j\phi_l(t)}.
\tag{2.37}
$$

In (2.37) it is clear that the time-variant channel given by $h_b(t, \tau)$ represents the response of the multipath channel at the instant $t$ to an impulse that stimulated the channel at time $t - \tau_l(t)$. Since for some channels it is more suitable to represent the received signal as a continuum of multipath delays

[PM06], the received signal can be written as

$$y(t) = Re\left\{\left[\int_{-\infty}^{\infty}\alpha_l(t,\tau)s_b(t-\tau_l)e^{-j\phi_l(t)}d\tau\right]e^{j2\pi f_c t}\right\} \qquad (2.38)$$

and the channel can be simplified to a time-varying complex amplitude related to the corresponding delay, i.e.,

$$h(t,\tau) = \alpha_l(t,\tau)e^{-j\phi_l(t)} \qquad (2.39)$$

where $h(t,\tau)$ gives the response of the channel at time $t$ due to an impulse applied at $(t-\tau)$ (in other words, $t$ indicates the instant in which the channel is used, while the parameter $\tau$ reflects the elapsed time since the input was applied, i.e., delay).

The equivalent lowpass channel $h_b(t,\tau)$ consists of the sum of a large number of attenuated, delayed, and phase rotated impulses. Since the fading effect is mainly due to the randomly time-variant phases $\phi_l(t)$, the multipath propagation model of (2.37) causes the signal to fade [PM06], as shown in Fig. 2.11. For instance, considering the carrier frequencies employed in the typical mobile communication systems, the $l$th multipath component will have a $f_c\tau_l(t) \gg 1$. Consider an indoor application with a carrier frequency of $f_c = 1$ GHz and $\tau_l = 50$ ns. In this case $f_c\tau_l(t) = 50 \gg 1$. Regarding outdoor systems, much greater values of multipath delays have to be considered, and therefore this property still applies. It is important to note that when $f_c\tau_l(t) \gg 1$ the small changes in the path delay $\tau_l$ will lead to a large phase change in the $l$ multipath overall phase $\phi_l(t)$, which means that the phase can be regarded as random and uniformly distributed. If in addition we consider that different scatterers are independent, applying the CLT we can assume that $h_b(t,\tau)$ is approximately a complex Gaussian random process.

Consider a mobile station moving at speed $v$, within a multipath propagation environment. As the mobile station moves its position changes as well as the characteristics of each propagation path. Assuming that the movement occurs at a constant velocity $v$, the distance between a previous position and a new one is a function of time, i.e., $d = vt$. The motion produces Doppler shifts on the several incoming received waves. Fig. 2.12 illustrates a mobile station moving at a constant speed $v$, and it moves by $d$ from the initial point to the new point (to simplify; a two dimensions model is presented, so that the angle of arrival is the corresponding azimuth). If the mobile antenna moves a short distance $\Delta l$, the $l^{th}$ incoming ray, with an angle of arrival of $\theta_l$ with respect to the instantaneous direction of motion, will experience a shift in phase. The difference in the path lengths from the base station to the mobile station is given by $\Delta l = d\cos(\theta_l) = v\Delta t\cos(\theta_l)$. It is then clear that the length of the $l^{th}$ path increases by $\Delta l$. And as a consequence, the phase offset in received signal due to the difference in path lengths will be

$$\Delta\phi_l = \frac{2\pi\Delta l}{\lambda_c} = \frac{2\pi v\Delta t\cos(\theta_l)}{\lambda_c} \qquad (2.40)$$

where $\lambda_c = \frac{c}{f_c}$ denotes the wavelength at the carrier frequency.

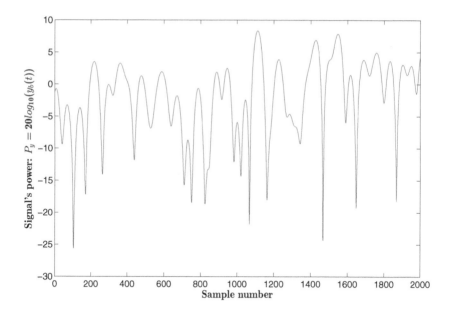

**Figure 2.11: Fast fading due to mobility: the signal strength exhibits a rapid variation with time.**

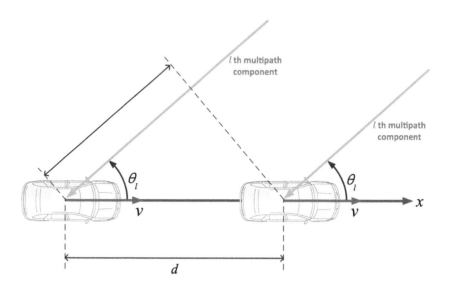

**Figure 2.12: Difference in path lengths from the transmitter to the mobile station.**

When modeling a mobile radio channel, it is common to make a set of assumptions about the propagation medium in order to simplify the process. Hence, in order to model the radio channel, we will use the model defined by Clarke [Cla98] and extended by Jakes [JC94], in which it is assumed that:

- the transmission occurs in the two-dimensional (horizontal) plane;

- the receiver is assumed to be located in the center of an isotropic scattering area;

- the angles of the waves arriving at the receiving antenna are given by $\theta$ and assumed to be uniformly distributed in the interval $[0; 2\pi]$;

- the antenna radiation pattern of the receiving antenna is a circular-symmetrical (omnidirectional antenna).

In sum, in this model the channel is assumed to consist of several scatterers disposed in a uniform scattering environment, closely situated in relation to the angle. This scenario is illustrated in Fig. 2.13 in which $L$ multipath components are placed in the uniform scattering environment with an angle of arrival $\theta_l = l\Delta\theta$, with $\Delta\theta$ corresponding to $\Delta\theta = 2\pi/L$.

The Doppler power spectral density $S(f)$ referring to the scattered components, is given by (assuming an omnidirectional antenna)

$$S(f) = \begin{cases} \frac{1}{\sqrt{1-(f/f_D)^2}}, & |f| \leq f_D, \\ 0 & |f| > f_D, \end{cases} \qquad (2.41)$$

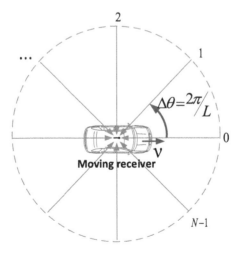

Figure 2.13: Example of an uniform scattering scenario.

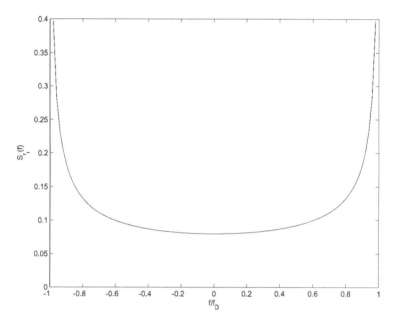

**Figure 2.14: Doppler power spectrum density.**

where $f_D$ stands for the maximum Doppler frequency. The result in (2.41) corresponds to the power spectral density originally proposed by Clarke, and later extended by Jakes, earning the name Doppler power spectral density (also referred to in the literature as Jakes power spectral density) [Cla98] [JC94]. This power spectrum density (PSD) is shown in Fig. 2.14 which illustrates the in-phase and quadrature power spectral densities, which tend to infinity when $f = \pm f_D$ (this means that the PSD is maximum at $\pm f_D$). However, since these notions are built on an approximation based on the uniform scattering environment, then this will not be true in practice, since this uniform scattering model, is not realistic. Nevertheless, in scenarios characterized by dense scattering, the PSD is typically maximized at frequencies near to $f_D$. The next step for the statistical description of the channel is to evaluate the time-correlation of the channel, in order to measure the degree of time-variation of the channel. Let us consider the random variables given by $h_b(t_1)$ and $h_b(t_2)$, assigned to the stochastic process $h_b(\tau)$ at the time instants $t_1$ and $t_2$, then,

$$r_{h_b h_b}(t_1, t_2) = E\left[h_b(t_1, \tau) h_b^*(t_2, \tau)\right]$$

$$= E\left[\sum_{l=0}^{L-1} |\alpha_l|^2 |\delta(\tau - \tau_l)|^2 e^{-j2\pi f_D (t_1 - t_2) \cos(\theta_l)}\right] \qquad (2.42)$$

where $r_{h_b h_b}(t_1, t_2)$ represents the autocorrelation function of $h_b(t)$. By assuming that not only the path amplitudes but also the delays are independent from the phases, while the phases are uniformly distributed between $[0, 2\pi]$, (2.42) can be simplified to

$$E\left[h_b(t_1, \tau) h_b^*(t_2, \tau)\right] = \frac{\sigma^2}{2\pi} \int_0^{2\pi} e^{-j2\pi f_D (t_1 - t_2) \cos(\theta_l)} \qquad (2.43)$$

and applying the integral representation of the zero*th*-order Bessel function of the first kind [Pat03] leads to

$$E\left[h_b(t_1, \tau) h_b^*(t_2, \tau)\right] = \sigma_T^2 J_0(2\pi f_D(t_1 - t_2)) \qquad (2.44)$$

in which $\sigma_T^2 = E\left[\sum_{l=0}^{L-1} |\alpha_l|^2\right]$ represents the total received power from the scatterers affecting the channel at given delay $\tau$.

Fig. 2.15 illustrates the zero*th*-order Bessel function of the first kind given by (2.44). From this plot it is possible to observe that the autocorrelation is zero for a value of $f_D \tau$ around $0.4\lambda$. By making $\nu\tau \approx 0.4$ we can make an important observation: considering the uniform scattering environment, the correlation is zero over a distance of approximately one $0.5\lambda$. However, the

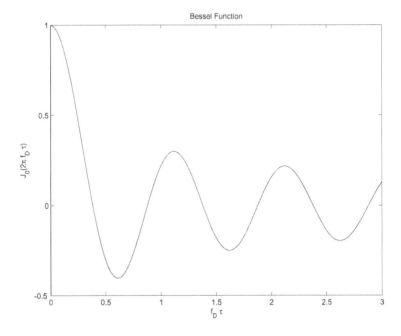

**Figure 2.15: Zeroth-order Bessel function of the first kind.**

signal still gets correlated after this, which means that for a distance greater than approximately one half wavelength the signal becomes independent from its initial value [JC94].

It is important to represent the time correlation of the channel in the discrete time. In terms of notation, the sample index is given by $n$ (not to be confused with the multipath component index)

$$E\left[h_{a,n}h_{b,n}^*\right] = \sigma_n^2 J_0(2\pi f_D T(a - b)) \tag{2.45}$$

A very important parameter arises from the previous equation. It is called normalized Doppler frequency and it is given by $f_D T$. It offers a comparison measure of the Doppler shift in relation to the carrier frequency, i.e.,

$$f_D T = \frac{f_c v}{c} T, \tag{2.46}$$

where $T$ denotes the symbol duration. The normalized Doppler is directly proportional to the motion speed and carrier frequency. Hence, in a real environment, the dynamics of a time varying channel can be described by $f_D T$.

# Chapter 3

# Block Transmission Techniques

This chapter starts with a brief introduction to multi-carrier (MC) and single-carrier (SC) modulations. It includes several aspects such as the analytical characterization of each modulation type, and some relevant properties of each modulation. For both modulations special attention is given to the characterization of the transmission and receiving structures, with particular emphasis on the transmitter and receiver performances. MC modulations and their relations with SC modulations are analyzed. Section 3.4.1 describes the OFDM modulation. Section 3.5 characterizes the basic aspects of the SC-FDE modulation including the linear and iterative FDE receivers. Finally, in Section 3.6, the performance of OFDM and SC-FDE for severely time-dispersive channels is compared.

## 3.1 Transmission Structure of a Multicarrier Modulation

An MC system transmits a multicarrier modulated symbol (composed of $N$ symbols on $N$ subcarriers in time $N/B$). First, a serial to parallel conversion is implemented in order to *demultiplex* the incoming high-speed serial stream and output several serial streams but of much lower speed. Subsequently, with resort to a constellation mapper, these parallel information bits are then modulated in the specified digital modulation format (phase shift keying (PSK), quadrature amplitude modulation (QAM), etc.). Posteriorly, each of the $N$

modulated symbols is associated onto the respective subcarrier with resort to a bank of $N$ sinusoidal oscillators, disposed in parallel, matched in frequency and phase to the $N$ orthogonal frequencies $(f_0, f_1, \ldots f_{N-1})$. Hence, each subcarrier is centered at frequencies that are orthogonal to each other. Finally, the signals modulated onto the $N$ subcarriers are summed forming the composite MC signal, which is then transmitted through the channel, as shown in Fig. 3.1.

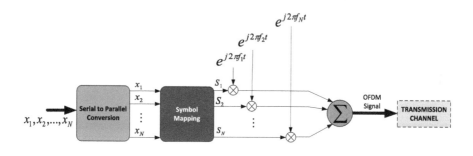

**Figure 3.1: Transmission structure for multicarrier modulation.**

MC modulation transmits a high-speed serial stream at the input, over several streams of lower data rate. As a consequence, the symbol period is extended, resulting in a significant advantage since the transmission becomes more resilient to the multipath environment. This is especially desirable in mobility scenarios, since it allows a reliable signal reception within fast-varying channels.

## 3.2 Receiver Structure of a Multicarrier Modulation

At the receiver, the received composite signal $y(t)$ is correlated with the set of subcarriers in a sort of a matched filtering operation (the matched filter uses a correlation process to detect the signal). The correlation of $y(t)$ with the $l^{th}$ coherent subcarrier[1] is a simple operation which can be expressed as

$$y(t)\left(e^{j2\pi f_l t}\right)^* = y(t)\left(e^{j2\pi l f_0 t}\right)^* \tag{3.1}$$

where $f_0 = \frac{B}{N}$ is the fundamental frequency. From Fourier series definition it can be inferred that all the other frequencies are in fact multiples of the fundamental frequency. Note that when recovering the symbols, the time period

---

[1] Coherent refers to equal in frequency and phase to the $k^{th}$ carrier.

of observation of the symbol (i.e., the detection window), corresponds to the time period of integration, which is mandatory to keep the orthogonality, and it consists of the fundamental period $T_0 = \frac{1}{f_0} = \frac{N}{B}$. Let us ignore the presence of noise and channel effects. Under these conditions, the received signal $y(t)$ equals the transmitted signal $s(t)$.

$$y(t) = s(t) = \sum_{k=0}^{N-1} S_k e^{j2\pi f_k t} \tag{3.2}$$

After taking the composite signal and correlating it with the corresponding $l^{th}$ coherent subcarrier, the result is integrated from 0 to the fundamental period $T_0$,

$$\frac{1}{T_0} \int_0^{T_0} y(t) \left(e^{j2\pi l f_0 t}\right)^* dt = \frac{1}{T_0} \int_0^{T_0} \left(\sum_{k=0}^{N-1} S_k e^{j2\pi k f_0 t}\right) e^{-j2\pi l f_0 t} dt, \tag{3.3}$$

which can simply be represented as (taking the summation out of the integral):

$$\sum_{k=0}^{N-1} \left[\frac{1}{T_0} \int_0^{T_0} S_k e^{j2\pi(k-l)f_0 t} dt\right]. \tag{3.4}$$

Let us focus on the integration term. The spacing of $f_0 = \frac{1}{T_0} = \frac{B}{N}$ between the subcarriers makes them orthogonal over each symbol period. This is a fundamental property expressed as

$$\frac{1}{T_0} \int_0^{T_0} \left(e^{j2\pi k f_0 t}\right) \left(e^{-j2\pi l f_0 t}\right) dt = \frac{1}{T_0} \int_0^{T_0} e^{j2\pi(k-l)\cdot f_0 t} dt = \begin{cases} 0, & \text{if } k \neq l; \\ T_0, & \text{if } k = l. \end{cases} \tag{3.5}$$

Coherent demodulation consists of correlating with $e^{(j2\pi f_l t)}$ and integrating over the fundamental period $T_0$. Hence, when we coherently demodulate the $l^{th}$ subcarrier, all subcarriers are orthogonal except the subcarrier corresponding to the $k^{th}$ symbol. In other words, if the above result is integrated over the fundamental period $T_0$, all the terms are zero except when $k = l$. Equation (3.4) can be rewritten as

$$\sum_{k=0}^{N-1} \left[\frac{1}{T_0} \int_0^{T_0} S_k e^{j2\pi(k-l)f_0 t} dt\right]$$

$$= \frac{B}{N} \int_0^{N/B} \left(S_l + \sum_{k \neq l} \left[S_k e^{(j2\pi(k-l)\frac{B}{N}t)}\right]\right) dt$$

$$= \frac{B}{N} S_l \frac{N}{B} + \frac{B}{N} \sum_{k \neq l} \left[S_k \int_0^{N/B} e^{(j2\pi(k-l)\frac{B}{N}t)} dt\right] \tag{3.6}$$

$$= \frac{B}{N} S_l \frac{N}{B} + 0 = S_l$$

From (3.6), it is clear that the information symbol $S_k$, transmitted by the $k^{th}$ subcarrier, can be recovered by coherently demodulating the composite signal at the receiver. This is done by locally generating the corresponding $l$ coherent subcarrier, equal in frequency and phase to the $k^{th}$ subcarrier, and then mix it with the received composite signal. The result is then integrated over the period $T_0$, and with this process, the respective symbol is recovered. After correlating the received composite signal with each of the $N$ different subcarriers, the $N$ detected information symbols are finally multiplexed into a serial stream through to a parallel to serial operation, as shown in Fig. 3.2.

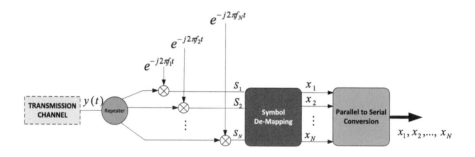

**Figure 3.2: Receiving structure for multicarrier modulation.**

## 3.3 Multicarrier Modulations or Single-Carrier Modulations?

In a conventional single carrier modulation, the energy of each symbol is distributed over the total transmission band. The term *single carrier* implies a unique carrier which occupies the entire communication bandwidth $B$, and the transmission is performed at a high symbol rate. Considering a bandwidth $B$, and assuming that one symbol is transmitted every $T$ seconds (in fact, two symbols can be transmitted on different sine and cosine carriers), then the symbol time is given by $T_S = \frac{1}{B}$. This leads to a symbol rate of $R = \frac{1}{T_S} = \frac{1}{1/B} = B$. For instance, if a bandwidth of 100 MHz is available, we can transmit symbols at a rate of 100 Mbps, employing a symbol time of $T_S = \frac{1}{B} = \frac{1}{100}$ MHz $= 10\mu s$. One might think that since an MC modulation scheme transmits $N$ symbols in parallel, it increases the throughput. However, the observation time also increases due to the fact that multicarrier modulation transmits $N$ symbols using $N$ subcarriers within the time period $T_S = N/B$, leading to a symbol rate of $R = \frac{N}{N/B} = B$. In comparison, a single-carrier scheme transmits one symbol in time period $T_S = 1/B$, with a rate of $R = \frac{1}{1/B} = B$. Obviously, $N$ symbols in $N/B$ time (in the case of an MC) or 1 symbol in $1/B$ time (in SC) are both the same with respect to the signal throughput, as illustrated in Fig. 3.3.

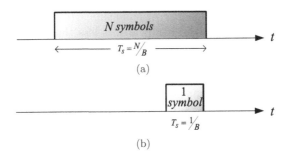

**Figure 3.3: (a) Transmission of $N$ information symbols on $N$ subcarriers in time $N/B$; (b) transmission of 1 information symbol in $1/B$ time.**

Previously, we have stated that the overall data rate is the same in multicarrier and single-carrier modulation schemes. We may think that when compared to the SC modulation, the MC modulation is just an extremely complicated system without advantage over an SC system (since both schemes have an overall data rate of $B$ symbols per second). So, if from the symbol rate perspective, the MC system and the SC system are equivalent, what advantages does the much more complex MC system have to offer?

In order to better understand the fundamental advantage of MC modulations, consider a scenario in which the available bandwidth for transmission is $B=1024$ kHz. An SC system will use the complete bandwidth of 1024 kHz, much greater than the coherence bandwidth of the channel (i.e., $B \geq B_C$), which is assumed to be approximately 200 to 300 kHz. In these conditions, since the bandwidth is much greater than the coherence bandwidth, the channel is said to be frequency selective (different frequency components of the signal experience different fading), which implies ISI in the time domain. Therefore, a high bit rate SC digital signal experiences frequency selective fading and ISI occurs, which may result in significant distortion since the symbols interfere with each other, highly distorting the received signal and affecting the reliable detection of the symbols. Now consider an MC system with the same available bandwidth for transmission but with $N = 256$ subcarriers. In this case, the bandwidth of each subcarrier is $\frac{B}{N} = \frac{1024}{256} = 4$ kHz, much less than the coherence bandwidth considered (i.e., $B_C \sim 200 - 300 \gg 4$ kHz). Each subcarrier will then experience frequency flat fading in the frequency domain, and no ISI in the time domain will occur. So, what initially was a wideband radio channel, was divided into several narrowband (ISI-free) subchannels for transmission in parallel.

In comparison with the SC scheme, the overall data rate remains unchanged. However, the much more complex implementation trade-off has a significant advantage: it is possible to implement a ISI free reliable detection scheme at the receiver side. The narrowband subcarriers experience flat fading in the frequency domain, as the bandwidth is less than the coherence bandwidth. Hence, the major motivation behind MC modulations was to convert a frequency selective wideband channel into a non-frequency selective channel. Nevertheless, it is important to note that if the implementation of a coherent modulator is significantly challenging, implementing a bank of $N$ modulators can get extremely complicated in hardware. Since the MC modulation requires a bank of $N$ modulators, proportional to the number of subcarriers. Hence, the modulation, coherent demodulation, and synchronization requirements of the MC modulation scheme led to a very complex system, very susceptible to loss of orthogonality and ICI.

## 3.4 OFDM Modulations

OFDM was initially proposed by R. Chang in 1966 [Cha66]. His work presented an approach for multiple transmission of signals over a band-limited channel, free of ISI. By dividing the frequency selective channel into several frequency narrowband channels, the smaller individual channels would be subjected to flat fading. Using the Fourier transform, Chang was able to provide a method to guarantee the orthogonality among the parallel channels (or subcarriers), through the summation of sine and cosine. The orthogonality between the subcarriers within an MC modulation is crucial since, as has been seen before, it allows parallel channel data transmission rates equivalent to the bandwidth of the channel, corresponding to half the ideal Nyquist rate. However, due to its complexity, Chang's system was still hard to implement. The Fourier transforms rely on oscillators whose phase and frequency have to be very precise.

Moreover, as was shown before, the complexity of the MC scheme requires a bank of $N$ modulators, proportional to the number of subcarriers. If the implementation of a coherent modulator is significantly challenging, implementing thousands of parallel subcarriers in hardware is extremely difficult, even with state-of-the-art technology. Hence, the modulation, coherent demodulation, and synchronization requirements led to a very complex OFDM analog system, known to be very susceptible to loss of orthogonality and ICI.

In the early 1970s, Weinstein and Ebert [WE71], proposed a technique that helped to solve the complexity problem of implementing the $N$ modulators and demodulators. With resort to the discrete Fourier transform (DFT), they proposed a method to digitally implement the baseband modulation and demodulation. This approach suppressed the bank modulators and demodulators, highly simplifying the implementation and at the same time ensuring the orthogonality between subcarriers. The DFT converts the information symbols from the time domain to the frequency domain, and the output result is a function of the sampling period $T_S$ and the number of sample points $N$. Each of the $N$ frequencies represented in the DFT is a multiple of the fundamental frequency $f_0 = \frac{1}{NT_S}$, where the sampling time is given by $T_S = \frac{1}{\text{sampling rate}} = \frac{1}{B}$, with the product $N \cdot T_S$ corresponding to the total sample time. In its turn, the dual function IDFT converts a signal defined by its frequency components to the corresponding time domain signal, with the duration $NT_S$. According to the well-known result from sampling theorem, a bandlimited signal can be fully reconstructed from the samples at the receiver, as long it is sampled at a rate twice the maximum frequency (Nyquist rate). In order to better understand this, we will take the MC signal, or the MC composite signal $y(t)$ defined by

$$y(t) = \sum_{k=0}^{N-1} S_k \cdot e^{j2\pi k \frac{B}{N} t} \tag{3.7}$$

and sample it at rate $B$. The $u^{th}$ sample is taken at

$$t = uT_S = \frac{u}{B}, \tag{3.8}$$

and therefore,

$$y(uT_S) = x(u) = \sum_k S_n \cdot e^{j2\pi n \frac{B}{N} \frac{u}{B}} = \sum_n S_n \cdot e^{j2\pi n \frac{u}{N}} \tag{3.9}$$

where the left term of the above equation, $x(u)$ represents the samples of the MC signal, while the right term, $\sum_n S_n \cdot e^{j2\pi n \frac{u}{N}}$ represents the discrete Fourier transform (DFT) of $S$; this is the DFT of the information symbol. So this powerful result by Weinstein and Ebert [WE71] shows that there is no need to use $N$ modulators and $N$ demodulators. This is very effective since in order to obtain the samples of the MC transmitted symbol, it is just needed to take the $N$ information symbols, and compute their DFT (assuming the absence of noise).

The processing time can be reduced with resort to the fast Fourier transform (FFT), and the inverse FFT (IFFT). The FFT is a key process to separate the carriers of an OFDM signal. It was developed by Cooley and Tukey [CT65], and it consists of a very fast algorithm for computing the DFT, capable of reducing the number of arithmetic operations by decreasing the number of complex multiplication operations from $N^2$ to $\frac{N}{2} log_2 N$, for an $N-$point IDFT or DFT (with $N$ representing the size of the FFT). This allows a much more practical Fourier analysis since it simply samples the analog composite signal with an analog-to-digital converter (ADC), submitting the resulting samples to the FFT process. The FFT operation at the receiver separates the signal components into the $N$ individual subcarriers and sorts all the signals to recreate the original data stream. On the other hand, the individual digital modulated subcarriers are submitted to the IFFT operation, which forms the composite signal to be transmitted. The IFFT is a conversion process from frequency domain into time domain, so the IFFT can be used at the transmitter to convert frequency domain samples to time domain samples, and hence generate the OFDM symbol.

The FFT is formally described as follows:

$$X(k) = \sum_{n=0}^{N-1} x(n) \sin\left(\frac{2\pi kn}{N}\right) + j \sum_{n=0}^{N-1} x(n) \cos\left(\frac{2\pi kn}{N}\right), \qquad (3.10)$$

where as its dual, IFFT is given by

$$x(n) = \sum_{n=0}^{N-1} X(k) \sin\left(\frac{2\pi kn}{N}\right) - j \sum_{n=0}^{N-1} X(k) \cos\left(\frac{2\pi kn}{N}\right). \qquad (3.11)$$

The equations of the FFT and IFFT differ the coefficients they take and the minus sign. Both equations do the same operation, i.e., multiply the incoming signal with a series of sinusoids and separate them into bins. In fact, FFT and IFFT are dual and behave in a similar way. Moreover, the IFFT and FFT blocks are interchangeable.

Fig. 3.4 illustrates how the use of the IFFT block in the transmitter avoids the need for separate sinusoidal converters (note that IFFT and FFT blocks in the transmitter are interchangeable as long as their duals are used in receiver).

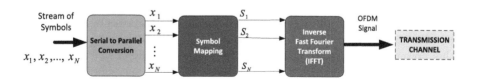

**Figure 3.4: Transmission structure for multicarrier modulation with re-sort to the IFFT block.**

### 3.4.1 Analytical Characterization of the OFDM Modulations

The complex envelope of an OFDM signal, given by (3.12), is characterized by a sum of blocks (also referred to as bursts), transmitted at a rate $F \geq \frac{1}{T_B}$. The duration of each block is $T_B \geq T$, in which $T = \frac{1}{F}$ denotes the duration of the payload part.

$$s(t) = \sum_m \left[ \sum_{k=0}^{N-1} S_k^{(m)} e^{j2\pi k F t} \right] r(t - mT_B), \tag{3.12}$$

where $S_k^{(m)}$ represents the OFDM symbol transmitted on the $k^{th}$ subcarrier of a given block $m$, in the frequency domain. Hence, the $N$ data symbols $\{S_k; k = 0, ..., N-1\}$ are sent during the $m^{th}$ block, with the group of complex sinusoids $\{e^{j2\pi k F t}; k = 0, ..., N-1\}$ denoting the $N$ subcarriers. Let us consider the $m^{th}$ OFDM block. During the OFDM block interval, the transmitted signal can be expressed as

$$s^{(m)}(t) = \sum_{k=0}^{N-1} S_k^{(m)} r(t) e^{j2\pi k F t} = \sum_{k=0}^{N-1} S_k^{(m)} r(t) e^{j2\pi \frac{k}{T} t}, \tag{3.13}$$

with the pulse shape, $r(t)$, defined as

$$r(t) = \begin{cases} 1, & [-T_G, T] \\ 0, & \text{otherwise}, \end{cases}$$

where $T = \frac{1}{F}$ and $T_G \geq 0$ corresponds to the duration of the "guard interval" used to compensate time-dispersive channels. Therefore $r(t)$ is a rectangular pulse, with a duration that should be greater than $T$ (i.e., $T_B = T + T_G \geq T = \frac{1}{F}$), to be able to deal with the time-dispersive characteristics of the channels. The subcarrier spacing $F = \frac{1}{T}$, guarantees the orthogonality between the subcarriers over the OFDM block interval. The different subcarriers are orthogonal during the interval $[0, T]$, which coincides with the effective detection interval, since

$$\int_0^T |r(t)|^2 e^{-j2\pi(k-k')Ft} dt = \int_0^T e^{-j2\pi(k-k')Ft} dt = \begin{cases} 1, & k = k', \\ 0, & k \neq k'. \end{cases} \tag{3.14}$$

Therefore, for each sampling instant, we may write (3.13) as

$$s^{(m)}(t) = \sum_{k=0}^{N-1} S_k e^{j2\pi k F t}, \quad 0 \leq t \leq T_B. \tag{3.15}$$

In spite of the overlap of the different subcarriers, the mutual influence among them can be avoided. Under these conditions, the bandwidth of each subcarrier becomes small when compared with the coherence bandwidth of the channel (i.e., the individual subcarriers experience flat fading, which allows simple equalization). This means that the symbol period of the subcarriers must be longer than the delay spread of the time-dispersive radio channel.

From (3.15), the $m^{th}$ block should take the form

$$s^{(m)}(t) = \sum_{k=0}^{N-1} S_k^{(m)} e^{j2\pi k F t} = \sum_{k=0}^{N-1} S_k^{(m)} e^{j2\pi \frac{k}{T_B} t} = \sum_{k=0}^{N-1} S_k^{(m)} e^{j2\pi f_k t},$$

$$0 \leq t \leq T_B, \tag{3.16}$$

where $\{S_k^{(m)}; k = 0, ..., N - 1\}$ represents the data symbols of the $m^{th}$ burst, $\{e^{j2\pi f_k t}; k = 0, ..., N - 1\}$ are the subcarriers, $f_k = \frac{k}{T_B}$ is the center frequency of the $k^{th}$ subcarrier. It is also assumed that $r(t) = 1$ in the interval $[-T_G, T]$.

By applying the inverse Fourier transform to both sides of (3.16), we obtain

$$S(f) = \mathcal{F}\{s(t)\} = \sum_{k=0}^{N-1} S_k^{(m)} sinc\left[\left(f - \frac{k}{T_B}\right)\right], \quad (3.17)$$

where the center frequency of the $k^{th}$ subcarrier is $f_k = \frac{k}{T_B}$, with a subcarrier spacing of $\frac{1}{T_B}$, that assures the orthogonality during the block interval (as stated by (3.14)).

Fig. 3.5 depicts the PSD of an OFDM signal, as well as the individual subcarrier spectral shapes for $N = 16$ subcarriers and data symbols. As we can see from Fig. 3.5, when the $k^{th}$ subcarrier PSD ($f_k = \frac{k}{T_B}$) has a maximum, the adjacent subcarriers have zero-crossings, which achieve null interference between carriers and improve the overall spectral efficiency.

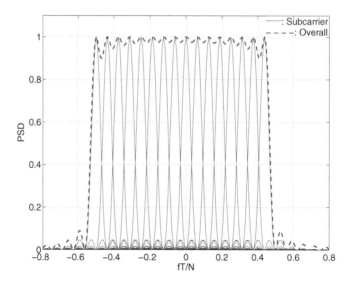

**Figure 3.5: The power density spectrum of the complex envelope of the OFDM signal, with the orthogonal overlapping subcarriers spectrum ($N = 16$).**

Since the duration of each symbol is long, a guard interval is inserted between the OFDM symbols to eliminate inter-block interference (IBI). If this guard interval is a cyclic prefix instead of a zero interval, it can be shown that inter-carrier interference (ICI) can also be avoided provided that only the useful part of the block is employed for detection purposes [Bin90]. Therefore, equation (3.16) is a periodic function in $t$, with period $T_B$, and the complex envelope associated with the guard period can be regarded as a repetition of the multicarrier blocks's final part, as exemplified in Fig. 3.6. Thus, it is valid to write

$$s(t) = s(t + T), \quad -T_G \leq t \leq 0. \tag{3.18}$$

Consequently, the guard interval is a copy of the final part of the OFDM symbol which is added to the beginning of the transmitted symbol, making the transmitted signal periodic. The cyclic prefix, transmitted during the guard interval, consists of the end of the OFDM symbol copied into the guard interval, and the main reason to do that is on the receiver that integrates over an integer number of sinusoid cycles each multipath when it performs OFDM demodulation with the FFT [CT65]. The guard interval also reduces the sensitivity to time synchronization problems.

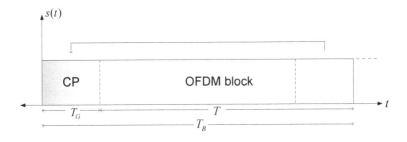

Figure 3.6: MC burst's final part repetition in the guard interval.

## 3.4.2 Transmission Structure

Let us now focus on the transmission of the OFDM signal where to simplify it is assumed a noiseless transmission case. Since it is an MC scheme, the incoming high data rate is split into $N$ streams of much lower rate by a serial/parallel converter. The parallel information bits are then modulated with a given digital modulation format, forming the symbols. The data is therefore transmitted by blocks of $N$ complex data symbols with $\{S_k; k = 0, ..., N - 1\}$ being chosen from a selected constellation (for example, a PSK constellation, or a QAM). The $N$ individual digital modulated symbols are then submitted to an IFFT operation in order to convert the frequency domain samples to time domain. The output corresponds to the OFDM symbol of (3.16), and if we sample the OFDM signal with an interval of $T_a = \frac{T}{N}$ we get the samples

$$s_n^{(m)} \equiv s(t)^{(m)}|_{t=nT_a} = s(t)\delta(t - nT_a) = \sum_{k=0}^{N-1} S_k^{(m)} e^{j2\pi \frac{k}{T} nT_a},$$

$$n = 0, 1, ..., N - 1, \tag{3.19}$$

where $F = \frac{1}{T}$. Consequently, (3.19) can be written as

$$s_n^{(m)} = \sum_{k=0}^{N-1} S_k^{(m)} e^{j\frac{2\pi kn}{N}} = IDFT\{S_k\}, \quad n = 0, 1, ..., N - 1. \tag{3.20}$$

Hence, referring to the $m^{th}$ block, $\{s_n^{(m)}; n = 0, ..., N-1\} = IDFT\{S_k^{(m)}; k = 0, ..., N - 1\}$. The IDFT operation can be implemented through an IFFT which is more computationally efficient. At the output of the IFFT, a cyclic prefix of $N_G$ samples, is inserted at the beginning of each block of $N$ IFFT coefficients. It consists of a time-domain cycle extension of the OFDM block, with size larger than the channel impulse response (i.e., the $N_G$ samples assure that the CP length is equal to or greater than the channel length). The cycle prefix is appended between each block, in order to transform the multipath linear convolution into a circular one. Thus, the transmitted block is $\{s_n; n = -N_G, ..., N - 1\}$, and the time duration of an OFDM symbol is $N_G + N$ times larger than the symbol of an SC modulation. Clearly, the CP is an overhead that costs power and bandwidth since it consists of additional redundant information data. Therefore, the resulting sampled sequence is described by

$$s_n^{(m)} = \sum_{k=0}^{N-1} S_k^{(m)} e^{j\frac{2\pi kn}{N}}, \quad n = -N_G, 1, ..., N - 1. \tag{3.21}$$

After a parallel to serial conversion, this sequence is applied to a digital-to-analog converter (DAC), whose output would be the signal $s(t)$. The signal is upconverted and sent through the channel. The resulting IDFT samples are then submitted to a digital-to-analog conversion operation performed by a DAC. Fig. 3.7 illustrates a simple OFDM transmission chain block diagram.

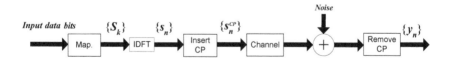

**Figure 3.7: Basic OFDM transmission chain.**

### 3.4.3 Reception Structure

At the channel output (after the RF down conversion), the received signal waveform $y(t)$ consists of the convolution of $s(t)$ with the channel impulse response, $h(\tau, t)$, plus the noise signal $n(t)$, i.e.,

$$y(t) = \int_{-\infty}^{+\infty} s(t - \tau)h(\tau, t)d\tau + n(t). \tag{3.22}$$

The received signal $y(t)$ is then submitted to an analog-to-digital converter (ADC), and sampled at a rate $T_a = \frac{T}{N}$. The resulting sequence $y_n$ consists of a set of $N+N_G$ samples, with the $N_G$ samples being extracted before the demodulation operation. The remaining samples $\{y_n; n = 0, ..., N - 1\}$ are demodulated through the DFT (performed by an FFT algorithm) to convert each block back to the frequency domain, followed by the baseband demodulation. For a given block, the resulting frequency domain signal $\{Y_k; k = 0, ..., N-1\}$, will be

$$Y_k = \sum_{k=0}^{N-1} y_n e^{-j\frac{2\pi kn}{N}}, \quad k = 0, 1, ..., N - 1. \tag{3.23}$$

The OFDM signal detection is based on signal samples spaced by a period of duration $T$. Due to multipath propagation, the received data bursts overlap leading to a possible loss of orthogonality between the subcarriers, as showed in Fig. 3.8(a). However, with resort to a CP of duration $T_G$ (greater than overall channel impulse response), the overlapping bursts in received samples during the useful interval are avoided, as shown in Fig. 3.8(b). Since IBI can be prevented through the CP inclusion, each subcarrier can be regarded individually.

The OFDM receiver structure is implemented employing an $N$ size DFT as shown in Fig. 3.9. Assuming flat fading on each subcarrier and null ISI, the received symbol is characterized in the frequency-domain by

$$Y_k = H_k S_k + N_k, \quad k = 0, 1, ..., N - 1, \tag{3.24}$$

where $H_k$ denotes the overall channel frequency response for the $k^{th}$ subcarrier and $N_k$ represents the additive Gaussian channel noise component.

On the other hand, the frequency-selective channel's effect, as the fading caused by multipath propagation, can be considered constant (flat) over an OFDM subcarrier if it has a narrow bandwidth (i.e., when the number of subchannels is sufficiently large). Under these conditions, the equalizer only has to multiply each detected subcarrier (each Fourier coefficient) by a constant complex number. This makes equalization far simpler at the OFDM receiver when compared to the conventional single-carrier modulation case. Additionally, from the computation's point of view, frequency-domain equalization is simpler than the corresponding time-domain equalization, since it only requires an FFT and a simple channel inversion operation. After acquiring the $Y_k$ samples, the data symbols are obtained by processing each one of

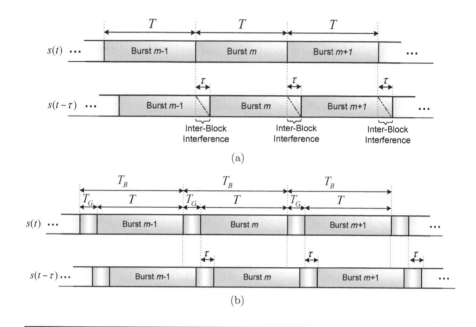

(a)

(b)

**Figure 3.8:** (a) Overlapping bursts due to multipath propagation; (b) IBI cancelation by implementing the cyclic prefix.

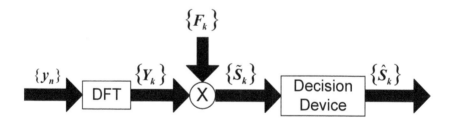

**Figure 3.9:** OFDM basic FDE structure block diagram with no space diversity.

the $N$ samples (in the frequency domain) with a frequency-domain equalization (FDE), followed by a decision device. Consequently, the FDE is a simple one-tap equalizer [PM06]. Hence, the channel distortion effects (for an uncoded OFDM transmission) can be compensated by the receiver depicted in Fig. 3.9, where the equalization process can be accomplished by an FDE optimized under the ZF criterion, with the equalized frequency-domain samples at the $k^{th}$ subcarrier given by

$$\tilde{S}_k = F_k Y_k. \tag{3.25}$$

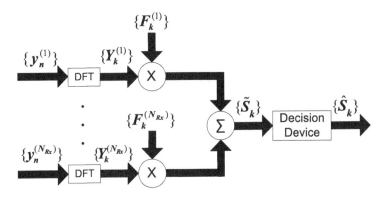

**Figure 3.10: OFDM receiver structure with a $N_{Rx}$-branch space diversity.**

In (3.25) $\tilde{S}_k$ represents the estimated data symbols which are acquired with the set of coefficients $\{F_k; = k = 0, 1, ..., N-1\}$, expressed by

$$F_k = \frac{1}{H_k} = \frac{H_k^*}{|H_k|^2}. \tag{3.26}$$

Naturally, the decision on the transmitted symbol in a subcarrier $k$ can be based on $\tilde{S}_k$.

Let us consider the case in which we have $N_{Rx}$-order space diversity. In Fig. 3.10 a maximal-ratio combining (MRC) [Kai95] diversity scheme is implemented for each subcarrier $k$. Therefore, the received sample for the $l^{th}$ receive antenna and the $k^{th}$ subcarrier is denoted by

$$Y_k^{(l)} = S_k H_k^{(l)} + N_k^{(l)}, \tag{3.27}$$

with $H_k^{(l)}$ denoting the overall channel frequency response between the transmit antenna and the $l^{th}$ receive antenna for the $k^{th}$ frequency, $S_k$ denoting the frequency-domain of the transmitted blocks, and $N_k^{(l)}$ denoting the corresponding channel noise. The set of equalized samples $\{\tilde{S}_k; k = 0, 1, \ldots, N-1\}$, are

$$\tilde{S}_k = \sum_{l=1}^{N_{Rx}} F_k^{(l)} Y_k^{(l)}, \tag{3.28}$$

where $\{F_k^{(l)}; k = 0, 1, \ldots, N-1\}$ is the set of FDE coefficients related to the $l^{th}$ diversity branch, denoted by

$$F_k^{(l)} = \frac{H_k^{(l)*}}{\sum\limits_{l'=1}^{N_{Rx}} \left| H_k^{(l')} \right|^2}. \tag{3.29}$$

Finally, by applying (3.27) and (3.29) to (3.28), the corresponding equalized samples can then be given by

$$\tilde{S}_k = S_k + \frac{\sum\limits_{l=1}^{N_{Rx}} H_k^{(l)*}}{\sum\limits_{l'=1}^{N_{Rx}} \left| H_k^{(l')} \right|^2} N_k^{(l)}. \tag{3.30}$$

## 3.5 SC-FDE Modulations

One drawback of the OFDM modulation is the high envelope fluctuations of transmitted signal. Consequently, these signals are more susceptible to nonlinear distortion effects, namely those associated with a nonlinear amplification at the transmitter, resulting in a low power efficiency. This major constraint is even worse in the uplink since more expensive amplifiers and higher power back-off are required at the mobile.

Instead, when an SC modulation is employed with the same constellation symbols, the envelope fluctuations of the transmitted signal will be much lower. Thus, SC modulations are especially adequate for the uplink transmission (i.e., transmission from the mobile terminal to the base station), allowing cheaper user terminals with more efficient high-power amplifiers. Nevertheless, if conventional SC modulations are employed in digital communications systems requiring transmission bit rates of Mbits/s, over severely time-dispersive channels, high signal distortion levels can arise. Therefore, the transmission bandwidth becomes much higher than the channels' coherence bandwidth. As a consequence, high-complexity receivers will be required to overcome this problem [PM06].

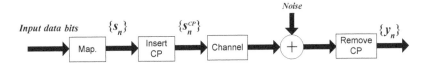

**Figure 3.11: Basic SC-FDE transmitter block diagram.**

### 3.5.1 Transmission Structure

In an SC-FDE modulation, data is transmitted in blocks of $N$ useful modulation symbols $\{s_n; n = 0, ..., N - 1\}$, resulting from a direct mapping of the original data into a selected signal constellation, for example QPSK. Posteriorly, a cyclic prefix with length longer than the channel impulse response is appended, resulting in the transmitted signal $\{s_n; n = -N_G, ..., N - 1\}$. The transmission structure of an SC-FDE scheme is shown in Fig. 3.11. As we can see the transmitter is quite simple since it does not implement a DFT/IDFT operation. The discrete versions of in-phase $(s_n^I)$ and quadrature $(s_n^Q)$ components are then converted by a DAC onto continuous signals $s^I(t)$ and $s^Q(t)$, which are then combined to generate the transmitted signal

$$s(t) = \sum_{n=-N_G}^{N-1} s_n r(t - nT_S), \tag{3.31}$$

where $r(t)$ is the support pulse and $T_S$ denotes the symbol period.

### 3.5.2 Receiving Structure

The received signal is sampled at the receiver and the CP samples are removed, leading in the time-domain the samples $\{y_n; n = 0, ..., N - 1\}$. As with OFDM modulations, after a size-$N$ DFT results in the corresponding frequency-domain block $\{Y_k; k = 0, ..., N - 1\}$, with $Y_k$ given by

$$Y_k = H_k S_k + N_k, \quad k = 0, 1, ..., N - 1, \tag{3.32}$$

where $H_k$ denotes the overall channel frequency response for the $k^{th}$ frequency of the block, and $N_k$ represents channel noise in the frequency-domain. The receiver structure is depicted in Fig. 3.12. After the equalizer, the frequency-domain samples referring to the $k^{th}$ subcarrier, $\tilde{S}_k$, are given by

$$\tilde{S}_k = F_k Y_k. \tag{3.33}$$

For a zero-forcing (ZF) equalizer the coefficients $F_k$ are given by (3.26), i.e.,

$$F_k = \frac{1}{H_k} = \frac{H_k^*}{|H_k|^2}. \tag{3.34}$$

From (3.34) and (3.32), we may write (3.33) as

$$\tilde{S}_k = F_k Y_k = \frac{Y_k}{H_k} = S_k + \frac{N_k}{H_k} = S_k + \epsilon_k. \tag{3.35}$$

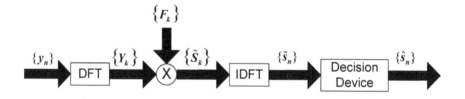

**Figure 3.12: Basic SC-FDE receiver block diagram.**

This means that the channel will be completely inverted. However, in the presence of a typical frequency-selective channel, deep notches in the channel frequency response will cause noise enhancement problems, and as a consequence, there can be a reduction of the signal-to-noise ratio (SNR). This can be avoided by the optimization of the $F_k$ coefficients under the MMSE criterion. Although the MMSE does not attempt to fully invert the channel effects in the presence of deep fades, the optimization of the $F_k$ coefficients under the MMSE criterion allows to minimize the combined effect of ISI and channel noise, allowing better performances. The mean-square error (MSE), in time-domain, can be described by

$$\Theta(k) = \frac{1}{N^2} \sum_{k=0}^{N-1} \Theta_k, \tag{3.36}$$

where

$$\Theta_k = E\left[\left|\tilde{S}_k - S_k\right|^2\right] = E\left[|Y_k F_k - S_k|^2\right]. \tag{3.37}$$

The minimization of $\Theta_k$ in order to $F_k$, requires the MSE minimization for each $k$, which corresponds to impose the following condition,

$$min_{F_k} \left( E \left[ |Y_k F_k - S_k|^2 \right] \right), \quad k = 0, 1, ..., N - 1, \tag{3.38}$$

which results in the set of optimized FDE coefficients $\{F_k; k = 0, 1, ..., N - 1\}$ [GDE03]

$$F_k = \frac{H_k^*}{\alpha + |H_k|^2}. \tag{3.39}$$

In (3.39) $\alpha$ denotes the inverse of the SNR, given by

$$\alpha = \frac{\sigma_N^2}{\sigma_S^2}, \tag{3.40}$$

where $\sigma_N^2 = \frac{E[|N_k|^2]}{2}$ stands for the variance of the real and imaginary parts of the channel noise components $\{N_k; k = 0, 1, ..., N - 1\}$, and $\sigma_S^2 = \frac{E[|S_k|^2]}{2}$ represents the variance of the real and imaginary parts of the data samples components $\{S_k; k = 0, 1, ..., N - 1\}$. The term $\alpha$ can be seen as a noise-dependent term that avoids noise enhancement effects for very low values of the channel frequency response. The equalized samples in the frequency-domain $\{\tilde{S}_k; k = 0, 1, ..., N - 1\}$, must be converted to the time-domain through an IDFT operation, and the decisions on the transmitted symbols are made upon the resulting equalized samples $\{\tilde{s}_n; n = 0, 1, ..., N - 1\}$.

It is possible to extend the SC-FDE receiver for space diversity scenarios. Fig. 3.13 shows an SC-FDE receiver structure with an $N_{Rx}$-branch space diversity, where an MRC combiner is applied to each subcarrier $k$.

Considering the $N_{Rx}$-order diversity receiver, the equalized samples at the FDE's output, are given by

$$\tilde{S}_k = \sum_{l=1}^{N_{Rx}} F_k^{(l)} Y_k^{(l)} \tag{3.41}$$

where $\{F_k^{(l)}; k = 0, 1, ..., N - 1\}$ is the set of FDE coefficients related to the $l^{th}$ diversity, which are given by

$$F_k^{(l)} = \frac{H_k^{(l)*}}{\alpha + \sum_{l'=1}^{N_{Rx}} \left| H_k^{(l')} \right|^2}, \tag{3.42}$$

with $\alpha = \frac{1}{SNR}$.

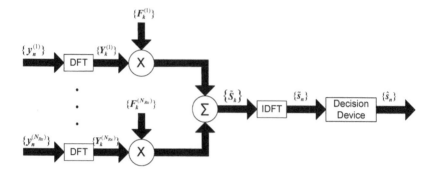

**Figure 3.13: Basic SC-FDE receiver block diagram with an $N_{Rx}$-order space diversity.**

## 3.6   Comparative Analysis between OFDM and SC-FDE

In order to compare OFDM and SC-FDE, refer to the transmission chains of both modulation systems, depicted in Fig. 3.14. Clearly, the transmission chains for OFDM and SC-FDE are essentially the same, except in the place where the IFFT operation is performed. In the OFDM, the IFFT is placed at the transmitter side to divide the data in different parallel subcarriers. For the SC-FDE, the IFFT is placed in the receiver to convert into the time-domain the symbols at the FDE output. Although there is lower complexity of the transmitter (it does not need the IDFT block), the SC-FDE requires a more complex receiver than OFDM. Consequently, from the point of view of over-all processing complexity (evaluated in terms of the number of DFT/IDFT blocks), both schemes are equivalent [SKJ94]. Moreover, for the same equal-

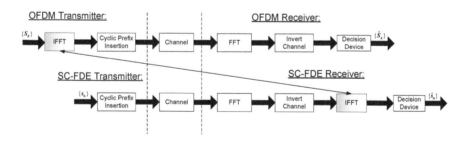

**Figure 3.14: Basic transmission chain for OFDM and SC-FDE.**

ization effort, SC-FDE schemes have better uncoded performance and lower envelope fluctuations than OFDM.

Fig. 3.15 presents a example of the performance results regarding uncoded OFDM modulations and uncoded SC-FDE modulations with ZF and MMSE equalization, for QPSK signals. The blocks are composed by $N = 256$ data symbols with a cycle prefix of 32 symbols. For simulation purposes, we consider a severely time dispersive channel with 32 equal power taps, with uncorrelated Rayleigh fading on each tap.

Without channel coding, the performance of the OFDM is very close to SC-FDE with ZF equalization. Moreover, SC-FDE has better uncoded performance under the same conditions of average power and complexity demands [GDCE00]. It should be noted that these results cannot be interpreted as if OFDM has poor performance, since the OFDM is severely affected by deep-faded subcarriers. Therefore, when combined with error correction codes, OFDM has a higher gain code when compared with SC-FDE [GDCE00]. Moreover, OFDM symbols are affected by strong envelope fluctuations and

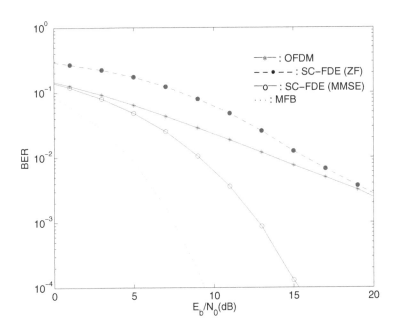

**Figure 3.15: Performance result for uncoded OFDM and SC-FDE.**

excessive peak-to-mean envelope power ratio (PMEPR) which causes difficulties related to power amplification and requires the use of linear amplification at the transmitter. On the other hand, the lower envelope fluctuation of SC signals allows a more efficient amplification. This is a very important aspect for the uplink transmission, where it is desirable to have low-cost and low-consumption power amplifiers. For downlink transmission, since the implementation complexity is gathered at the base stations where the costs and high power consumption are not major constraints, the OFDM schemes are a good option. Considering that both schemes are compatible, it is possible to have a dual-mode system where the user terminal employs an SC-FDE transmitter and an OFDM receiver, while the base station employs an OFDM transmitter and an SC-FDE receiver. Obviously, from Fig. 3.14, it becomes clear that this approach allows very low complexity mobile terminals the simpler SC transmissions and MC reception schemes.

## 3.7 DFE Iterative Receivers

Previously, it was shown that block transmission techniques, with appropriate cyclic prefixes and employing FDE techniques, are suitable for high data rate transmission over severely time dispersive channels. Typically, the receiver for SC-FDE schemes is a linear FDE, however, it is well known that nonlinear equalizers outperform linear ones [PM06] [BT02] [DGE03]. Among nonlinear equalizers the decision feedback equalizer (DFE), is a popular choice since it provides a good tradeoff between complexity and performance. Clearly, the previously described SC-FDE receiver is a linear FDE. Therefore, it would be desirable to design nonlinear FDEs, namely a DFE FDE. An efficient way of doing this is by replacing the linear FDE by an IB-DFE. IB-DFE is a promising iterative FDE technique, for SC-FDE. The IB-DFE receiver can be envisaged as an iterative FDE receiver where the feedforward and the feedback operations are implemented in the frequency domain. Due to the iteration process it tends to offer higher performance than a non-iterative receiver. These receivers can be regarded as low-complexity turbo FDE schemes [TH00], [TH01], where the channel decoder is not involved in the feedback. True turbo FDE schemes can also be designed based on the IB-DFE concept [BT05], [GTDE07]. In this section, we present a detailed study on schemes employing iterative frequency domain equalization.

### 3.7.1 IB-DFE Receiver Structure

Although a linear FDE leads to good performance for OFDM schemes, the performance of SC-FDE can be improved if the linear FDE is replaced by an IB-DFE [BT02]. The receiver structure is depicted in Fig. 3.16 [BDFT10, DGE03].

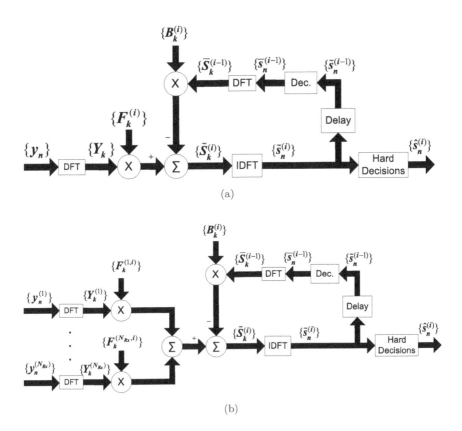

**Figure 3.16: IB-DFE receiver structure (a) without diversity (b) with a $N_{Rx}$-branch space diversity.**

In the case where an $N_{Rx}$-order space diversity IB-DFE receiver is considered, for the $i^{th}$ iteration, the frequency-domain block at the output of the equalizer is $\{\tilde{S}_k^{(i)}; k = 0, 1, \ldots, N-1\}$, with

$$\tilde{S}_k^{(i)} = \sum_{l=1}^{N_{Rx}} F_k^{(l,i)} Y_k^{(l)} - B_k^{(i)} \hat{S}_k^{(i-1)}, \qquad (3.43)$$

where $\{F_k^{(l,i)}; k = 0, 1, \ldots, N-1\}$ are the feedforward coefficients associated with the $l^{th}$ diversity antenna and $\{B_k^{(i)}; k = 0, 1, \ldots, N-1\}$ are the feedback coefficients. $\{\hat{S}_k^{(i-1)}; k = 0, 1, \ldots, N-1\}$ is the DFT of the block $\{\hat{s}_n^{(i-1)}; n = 0, 1, \ldots, N-1\}$, with $\hat{s}_n$ denoting the "hard decision" of $s_n$ from the previous FDE iteration. Considering an IB-DFE with "hard decisions," it can be shown that the optimum coefficients $B_k$ and $F_k$ that maximize the overall SNR, associated with the samples $\tilde{S}_k$, are [DGE03]

$$B_k^{(i)} = \rho \left( \sum_{l=1}^{N_{Rx}} F_k^{(l,i)} H_k^{(l)} - 1 \right), \tag{3.44}$$

and

$$F_k^{(l,i)} = \frac{\kappa H_k^{(l)*}}{\alpha + \left(1 - \left(\rho^{(i-1)}\right)^2\right) \sum_{l'=1}^{N_{Rx}} \left|H_k^{(l')}\right|^2}, \tag{3.45}$$

respectively, where $\rho$ denotes the so-called correlation factor, $\alpha = E[|N_k^{(l)}|^2]/E[|S_k|^2]$ (which is common to all data blocks and diversity branches), and $\kappa$ selected to guarantee that

$$\frac{1}{N} \sum_{k=0}^{N-1} \sum_{l=1}^{N_{Rx}} F_k^{(l,i)} H_k^{(l)} = 1. \tag{3.46}$$

Although the term *"IB-DFE with hard decisions"* is often referenced, the term *"IB-DFE with blockwise soft decisions"* would probably be more adequate, as we will see in the following. It can be seen from (3.44) and (3.45), that the correlation factor $\rho^{(i-1)}$ is a key parameter for the good performance of IB-DFE receivers, since it gives a blockwise reliability measure of the estimates employed in the feedback loop (associated with the previous iteration). This is done in the feedback loop by taking into account the hard decisions for each block plus the overall block reliability, which reduces error propagation problems. The correlation factor $\rho^{(i-1)}$ is defined as

$$\rho^{(i-1)} = \frac{E[\hat{s}_n^{(i-1)} s_n^*]}{E[|s_n|^2]} = \frac{E[\hat{S}_k^{(i-1)} S_k^*]}{E[|S_k|^2]}, \tag{3.47}$$

where the block $\{\hat{s}_n^{(i-1)}; n = 0, 1, \ldots, N-1\}$ denotes the data estimates associated with the previous iteration, i.e., the hard decisions associated with the time-domain block at the output of the FDE, $\{\tilde{s}_n^{(i)}; n = 0, 1, \ldots, N-1\}$ = IDFT $\{\tilde{S}_k^{(i)}; k = 0, 1, \ldots, N-1\}$.

For the first iteration, no information exists about $s_n$, which means that $\rho = 0$, $B_k^{(0)} = 0$, and $F_k^{(0)}$ coefficients are given by (3.39) (in this situation the IB-DFE receiver is reduced to a linear FDE). After the first iteration, the feedback coefficients can be applied to reduce a major part of the residual interference (considering that the residual bit error rate (BER) does not assume a high value). After several iterations and for a moderate-to-high SNR, the correlation coefficient will be $\rho \approx 1$ and the residual ISI will be almost totally canceled. In Fig. 3.17 is shown the average BER performance evolution for a fading channel. It refers to a transmission system with SC uncoded modulation, employing an IB-DFE receiver with 1, 2, 3, and 4 iterations. For comparative purposes, the corresponding performances of the MFB and additive white Gaussian noise (AWGN) channel are also included.

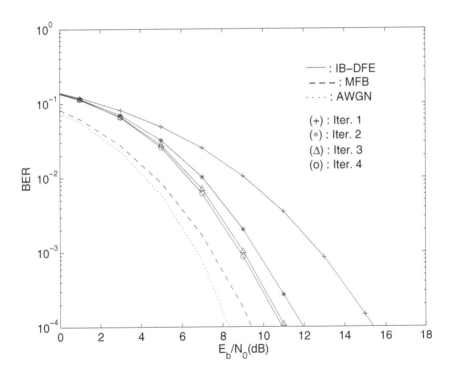

**Figure 3.17: Uncoded BER perfomance for an IB-DFE receiver with four iterations.**

From the results, we can see that the $E_b/N_0$ required for BER=$10^{-4}$ is around 15.5 dB for the $1^{st}$ iteration (that corresponds to the linear SC-FDE), decreasing to 11 dB after only three iterations, being clear that the use of the iterative receiver allows a significant performance improvement. Also, the asymptotic BER performance becomes close to the MFB after a few iterations.

It should be noted that (3.43) can be written as

$$\tilde{S}_k^{(i)} = \sum_{l=1}^{N_{Rx}} F_k^{(l,i)} Y_k^{(l)} - B_k^{'(i)} \overline{S}_{k,Block}^{(i-1)}, \tag{3.48}$$

where $B_k^{'(i)} = B_k^{(i)}/\rho^{(i-1)}$ and $\overline{S}_{k,Block}^{(i-1)} = \rho^{(i-1)} \hat{S}_k^{(i-1)}$ (as stated before, $\rho^{(i-1)}$ can be considered as the blockwise reliability of the estimates $\{\hat{S}_k^{(i-1)}; k = 0, 1, \ldots, N-1\}$).

### 3.7.2 IB-DFE with Soft Decisions

To improve the IB-DFE performance it is possible to use "soft decisions," $\bar{s}_n^{(i)}$, instead of "hard decisions," $\hat{s}_n^{(i)}$. Under these conditions, the "blockwise average" is replaced by "symbol averages" [GTDE07]. This can be done by using $\{\overline{S}_{k,Symbol}^{(i-1)}; k = 0, 1, \ldots, N-1\} = \text{DFT } \{\bar{s}_{n,Symbol}^{(i-1)}; n = 0, 1, \ldots, N-1\}$ instead of $\{\overline{S}_{k,Block}^{(i-1)}; k = 0, 1, \ldots, N-1\} = \text{DFT } \{\bar{s}_{n,Block}^{(i-1)}; n = 0, 1, \ldots, N-1\}$, where $\bar{s}_{n,Symbol}^{(i-1)}$ denotes the average symbol values conditioned to the FDE output from the previous iteration, $\hat{s}_n^{(i-1)}$. To simplify the notation, $\bar{s}_{n,Symbol}^{(i-1)}$ is replaced by $\bar{s}_n^{(i-1)}$ in the following equations.

For QPSK constellations, the conditional expectations associated with the data symbols for the $i^{th}$ iteration are given by

$$\bar{s}_n^{(i)} = \tanh\left(\frac{L_n^{I(i)}}{2}\right) + j \tanh\left(\frac{L_n^{Q(i)}}{2}\right) = \rho_n^I \hat{s}_n^I + j\rho_n^Q \hat{s}_n^Q, \tag{3.49}$$

with the log-likelihood ratio (LLR) of the "in-phase bit" and the "quadrature bit," associated with $s_n^I$ and $s_n^Q$, respectively, given by $L_n^{I(i)} = \frac{2}{\sigma_i^2} \tilde{s}_n^{I(i)}$ and $L_n^{Q(i)} = \frac{2}{\sigma_i^2} \tilde{s}_n^{Q(i)}$, respectively, with

$$\sigma_i^2 = \frac{1}{2} E[|s_n - \tilde{s}_n^{(i)}|^2] \approx \frac{1}{2N} \sum_{n=0}^{N-1} |\hat{s}_n^{(i)} - \tilde{s}_n^{(i)}|^2, \tag{3.50}$$

where the signs of $L_n^I$ and $L_n^Q$ define the hard decisions $\hat{s}_n^I = \pm 1$ and $\hat{s}_n^Q = \pm 1$, respectively. In (3.49), $\rho_n^I$ and $\rho_n^Q$ denote the reliabilities related to the "in-phase bit" and the "quadrature bit" of the $n^{th}$ symbol, and are given by

$$\rho_n^{I(i)} = E[s_n^I \hat{s}_n^I]/E[|s_n^I|^2] = \left|\tanh\left(\frac{L_n^{I(i)}}{2}\right)\right| \tag{3.51}$$

and

$$\rho_n^{Q(i)} = E[s_n^Q \hat{s}_n^Q]/E[|s_n^Q|^2] = \left| \tanh\left(\frac{L_n^{Q(i)}}{2}\right) \right|, \qquad (3.52)$$

respectively. Therefore, the correlation coefficient employed in the feedforward coefficients will be given by

$$\rho^{(i)} = \frac{1}{2N} \sum_{n=0}^{N-1} (\rho_n^{I(i)} + \rho_n^{Q(i)}). \qquad (3.53)$$

Obviously, for the first iteration $\rho_n^{I(0)} = \rho_n^{Q(0)} = 0$ and, consequently, $\bar{s}_n = 0$. Therefore, the receiver with "blockwise reliabilities" (hard decisions) and the receiver with "symbol reliabilities" (soft decisions) employ the same feedforward coefficients; however, in the first the feedback loop uses the "hard-decisions" on each data block, weighted by a common reliability factor, whereas in the second the reliability factor changes from bit to bit. From the performance results shown in Fig. 3.18, we observe clear BER improvements when we employ "soft decisions" instead of "hard decisions" in IB-DFE receivers.

The IB-DFE receiver can be implemented in two different ways, depending

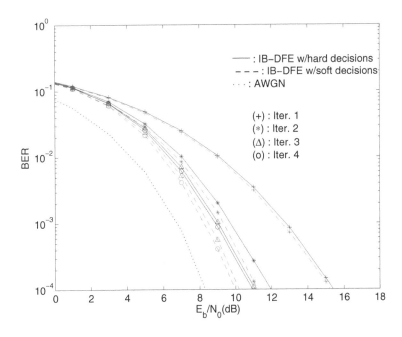

Figure 3.18: **Improvements in uncoded BER performance accomplished by employing "soft decisions" in an IB-DFE receiver with four iterations.**

on whether the channel decoding output is outside or inside the feedback loop. In the first case the channel decoding is not performed in the feedback loop, and this receiver can be regarded as a low complexity turbo equalizer implemented in the frequency domain. Since this is not a true "turbo" scheme, we will call it "conventional IB-DFE." In the second case the IB-DFE can be regarded as a turbo equalizer implemented in the frequency domain and therefore we will denote it as "turbo IB-DFE." For uncoded scenarios it only makes sense to employ conventional IB-DFE schemes. However, it is important to point out that in coded scenarios we could still employ a "conventional IB-DFE" and perform the channel decoding procedure after all the iterations of the IB-DFE. However, since the gains associated with the iterations are very low at low-to-moderate SNR values, it is preferable to involve the channel decoder in the feedback loop, i.e., to use the "turbo IB-DFE."

### 3.7.3 Turbo FDE Receiver

The most common way to perform detection in digital transmission systems with channel coding is to consider separately the channel equalization and channel decoding operations. However, using a different approach in which both operations are executed in conjunction, it is possible to achieve better performance results. This can be done employing turbo-equalization systems where channel equalization and channel decoding processes are repeated in an iterative way, with "soft decisions" being traversed through them. Turbo equalizers were firstly proposed for time-domain receivers. However, turbo equalizers can be implemented in the frequency-domain that, as conventional turbo equalizers, use "soft decisions" from the channel decoder output in the feedback loop.

The main difference between "conventional IB-DFE" and "turbo IB-DFE" is in the decision device: in the first case the decision device is a symbol-by-symbol soft-decision (for the QPSK constellation this corresponds to the hyperbolic tangent, as in (3.49)); for the turbo IB-DFE a SISO channel decoder (soft-in, soft-out) is employed in the feedback loop. The SISO block can be implemented as defined in [VY02], and provides the LLRs of both the "information bits" and the "coded bits." The input of the SISO block are the LLRs of the "coded bits" at the FDE output, given by $L_n^{I(i)}$ and $L_n^{Q(i)}$. It should be noted that the data bits must be encoded, interleaved, and mapped into symbols before transmission. The receiver scheme is illustrated in Fig. 3.19.

**Figure 3.19: SISO channel decoder for soft decisions.**

At the receiver side the equalized samples are demapped by a soft demapper followed by a deinterleaver providing the LLRs of the "coded bits" to the SISO channel decoder. The SISO operation is proceeded by a interleaver and after that a soft mapper provides the desired "soft decisions."

# Chapter 4

# Approaching the Matched Filter Bound

In the previous chapter it was shown that OFDM and SC-FDE block transmissions combined with frequency-domain detection schemes have been shown to be suitable for high data rate transmission over severely time-dispersive channels. Both modulations employ FDE at the receiver, whose implementation can be very efficient since the DFT/IDFT operations can be performed using the FFT algorithm. The receiver complexity is almost independent of the channel impulse response, making them suitable for severely time-dispersive channels [GDCE00, FABSE02]. Due to the lower envelope fluctuations of the transmitted signals, SC-FDE schemes are especially appropriate for the uplink transmission (i.e., the transmission from the mobile terminal to the base station), and OFDM schemes are preferable for the downlink transmission due to lower signal processing requirements at the receivers [GDCE00, FABSE02]. It has been shown that the IB-DFE, an iterative FDE technique for SC-FDE, can be regarded as low-complexity turbo FDE schemes where the channel decoder is not involved in the feedback.

Although the performance evaluation of these systems has been studied in several papers, the conditions for which the performance can be very close to the MFB for some channels, were not studied yet. In [DGE03], it was observed that the asymptotic performance of IB-DFE schemes can be sometimes very close to the MFB, but in other cases it is relatively far from it. Hence, it was not clear under which circumstances the performance can be close to the MFB. Therefore, the major motivation behind this study is to investigate how the performance of these systems can approach the MFB, in both uncoded and coded scenarios. For uncoded scenarios it is possible to present analytical

MFB results. On the other hand, for coded scenarios the situation is much more difficult due to the lack of closed formulas for the BER performance. For this reason, the performance study was done with resort to simulations (in this case with convolutional codes, but that also applies to other coding schemes). This work is concentrated on Rayleigh fading channels, which are widely used and where it is possible to obtain analytical MFB formulas for the uncoded case. For comparison purposes, a study concerning Nakagami channels is also included. Essentially, the conclusions drawn for Rayleigh fading scenarios are valid for other fading models, especially when the number of relevant multipath components is moderate or high.

This chapter presents a study on the impact of the number of multipath components, and the diversity order, on the asymptotic performance of OFDM and SC-FDE, in different channel coding schemes. It is shown that for a high number of separable multipath components the asymptotic performance of both schemes approaches the MFB, even without diversity. With diversity the performance approaches the MFB faster, with a small number of separable multipath components. It was also observed that the SC-FDE has an overall performance advantage over the OFDM option, especially when employing the IB-DFE with turbo equalization and for high code rates.

## 4.1   Matched Filter Bound

It is well known that the maximum likelihood receiver (MLR) represents the best possible receiver since it minimizes the probability of erroneous detection of a transmitted symbol. Considering that a sequence of data symbols is transmitted from a single source, and assuming the existence of ISI and Gaussian noise, the MLR consists of a matched filter followed by a sampler and a maximum likelihood sequence estimator implemented with the Viterbi algorithm [PM06] [For73]. If diversity is employed, the MLR consists of a bank of matched filters, one for each source, for each diversity branch. In this case, the outputs corresponding to each source being summed over all diversity branches. It is then followed by the bank of samplers (one sampler corresponding to each source), and a vector form of the maximum likelihood sequence estimator. It is difficult to analyze the performance of a maximum likelihood receiver due to its complexity, and as a consequence the exact calculation of the bit error probability of MLSE is difficult to accomplish. However, for uncoded transmission, a limit on the best attainable performance of a receiver operating in fading channels is given by the probability of error obtained assuming perfect equalization (i.e., the bit-error-rate achieved when the equalizer is capable of canceling all interference components). It consists of a theoretically optimal performance (not achievable in practice), and is called the matched filter bound [LSW81]. Therefore, the MFB represents the best possible error performance for a given receiver, and is obtained by assuming that just one symbol is transmitted, hence interference from neighboring symbols is avoided. As a consequence, there is no ISI, only additive noise. In these conditions, the optimum ML receiver is composed of a filter matched to h[k] and a decision can be made on the basis of the matched filter output at time k= 0.

### 4.1.1    *Approaching the Matched Filter Bound*

The MFB performance is defined as

$$P_{b,MFB} = E\left[Q\left(\sqrt{\frac{2E_b}{N_0}\frac{1}{N}\sum_{k=0}^{N-1}|H_k|^2}\right)\right], \tag{4.1}$$

and for an $N_{Rx}$-order space diversity, is written as

$$P_{b,MFB} = E\left[Q\left(\sqrt{\frac{2E_b}{N_0}\frac{1}{N}\sum_{k=0}^{N-1}\sum_{l=1}^{N_{Rx}}\left|H_k^{(l)}\right|^2}\right)\right], \tag{4.2}$$

where the expectation is over the set of channel realizations (it is assumed that $E[|H_k^{(l)}|^2 = 1])$.

Now, for the particular case in which a single ray is transmitted between the transmitter and each receiver antenna, the channel is known to exhibit a Rayleigh flat fading with the performance being given by [PM06]

$$P_{b,Ray} = \left(\frac{1-\mu}{2}\right)^{N_{Rx}}\sum_{l=0}^{N_{Rx}-1}\binom{N_{Rx}-1+l}{l}\left(\frac{1+\mu}{2}\right)^l, \tag{4.3}$$

with

$$\mu = \sqrt{\frac{\frac{E_b}{N_0}}{1+\frac{E_b}{N_0}}}. \tag{4.4}$$

However, for the general case in which different rays with different powers are considered, the calculation of the MFB is more complex. The analytical expressions for obtaining the MFB in uncoded scenarios when we have multipath propagation and diversity are presented in the following.

### 4.1.2    *Analytical Computation of the MFB*

Here is presented an analytical approach to obtain the MFB using an approach similar to [Lin95]. The analytical computation of the MFB only applies to the uncoded case. For the coded case it is very difficult to obtain analytical BER expressions (even for an ideal AWGN channel), since there are not closed formulas for the BER performance, and as a consequence the MFB needs to be computed by simulation. Consider the case of a transmission over a multipath Rayleigh fading channel with $N_{Rx}$ diversity branches, where all branches can have different fading powers or can be correlated. Assuming a discrete multipath channel for each diversity branch $l$, composed of $U_l$ discrete taps, where the magnitude of each tap $i$ has a mean square value of $\Omega_{i,l}^2$, the channel response, at time $t$, to a pulse applied at $t$-$\tau$, can be modeled as

$$c_l(\tau, t) = \sum_{i=1}^{U_l} \alpha_{i,l}(t)\,\delta(\tau - \tau_{i,l}), \quad l = 1...N_{Rx}, \tag{4.5}$$

with $\alpha_{i,l}(t)$ being a zero-mean complex Gaussian random process, $\tau_{i,l}$ the respective delay (assumed constant) and $\delta(t)$ is the Dirac function. For the derivation of the MFB, it is assumed the transmission of one pulse $s \cdot g(t)$, where $s$ is a symbol of a QPSK constellation and $g(t)$ is the impulse response of the transmit filter. Assuming a slowly time-varying channel, the sum of the sampled outputs, from the matched filters of the diversity branches, can be written as

$$y(t = t_0) = s \cdot \sum_{l=1}^{N_{Rx}} \sum_{i=1}^{U_l} \sum_{i'=1}^{U_l} \alpha_{i,l}\alpha_{i',l}^* R(\tau_{i,l} - \tau_{i',l}) + \sum_{l=1}^{N_{Rx}} \nu_l, \tag{4.6}$$

where $\nu_l$ represents AWGN samples with power spectral density $N_0$ and $R(\tau)$ is the autocorrelation function of the transmit filter. The instantaneous SNR is given by $SNR = \frac{2E_b}{N_0}\kappa$, where $E_b$ denotes the average bit energy and $\kappa$ is defined as

$$\kappa = \sum_{l=1}^{N_{Rx}} \sum_{i=1}^{U_l} \sum_{i'=1}^{U_l} \alpha_{i,l}\alpha_{i',l}^* R(\tau_{i,l} - \tau_{i',l}) = \mathbf{z}^H \mathbf{\Sigma} \mathbf{z}. \tag{4.7}$$

In (4.7), $\mathbf{z}$ represents a $U_{total} \times 1$ (with $U_{total} = \sum_{l=1}^{N_{Rx}} U_l$) vector containing the random variables $\alpha_{i,l}$ and $\mathbf{z}^H$ denotes the conjugate transpose of $\mathbf{z}$. $\mathbf{\Sigma}$ is a $U_{total} \times U_{total}$ Hermitian matrix constructed as

$$\mathbf{\Sigma} = \begin{bmatrix} \mathbf{R}_1 & \cdots & \mathbf{0} \\ \vdots & \ddots & \vdots \\ \mathbf{0} & \cdots & \mathbf{R}_{N_{Rx}} \end{bmatrix}, \tag{4.8}$$

where $\mathbf{R}_l$ is a matrix associated with the $l^{th}$ diversity branch, defined as

$$\mathbf{R}_l = \begin{bmatrix} R(0) & \cdots & R(\tau_{U_l,l} - \tau_{1,l}) \\ \vdots & \ddots & \vdots \\ R(\tau_{1,l} - \tau_{U_l,l}) & \cdots & R(0) \end{bmatrix}. \tag{4.9}$$

For a QPSK constellation the instantaneous BER can be written as

$$P_b\left(\kappa\right) = \frac{1}{2}\mathrm{erfc}\left(\sqrt{\frac{E_b}{N_0}\kappa}\right), \tag{4.10}$$

where $\mathrm{erfc}(x)$ is the complementary error function. The probability density function (PDF) of $\kappa$ can be obtained by writing it as a sum of uncorrelated random variables with known PDFs. Denoting $\boldsymbol{\Psi}$ as the covariance matrix of $\mathbf{z}$ ($\boldsymbol{\Psi} = Cov\left[\mathbf{z}\right]$), which is Hermitian and positive-semidefinite, it is possible to decompose $\boldsymbol{\Psi}$ into $\boldsymbol{\Psi} = \mathbf{Q}\mathbf{Q}^H$. In fact, by applying the Cholesky decomposition, $\mathbf{Q}$ will be a lower triangular matrix. Moreover, using this matrix, a new vector $\mathbf{z}' = \mathbf{Q}^{-1}\mathbf{z}$ can be defined, whose components will be uncorrelated unit-variance complex Gaussian variables and $\kappa$ becomes

$$\kappa = \mathbf{z}'^H \mathbf{Q}^H \boldsymbol{\Sigma} \mathbf{Q}\mathbf{z}' = \mathbf{z}'^H \boldsymbol{\Sigma}'\mathbf{z}', \tag{4.11}$$

with

$$\boldsymbol{\Sigma}' = \mathbf{Q}^H \boldsymbol{\Sigma} \mathbf{Q} = \boldsymbol{\Phi}\boldsymbol{\Lambda}\boldsymbol{\Phi}^H, \tag{4.12}$$

where $\boldsymbol{\Lambda}$ is a diagonal matrix whose elements are the eigenvalues $\lambda_i$ ($i=1,..,U_{total}$) of $\boldsymbol{\Sigma}'$ and $\boldsymbol{\Phi}$ is a matrix whose columns are the orthogonal eigenvectors of $\boldsymbol{\Sigma}'$. The decomposition of $\boldsymbol{\Sigma}'$ in (4.12) is possible due to its Hermitian property. Hence, (4.11) can be written as

$$\kappa = \mathbf{z}'^H \boldsymbol{\Phi}\boldsymbol{\Lambda}\boldsymbol{\Phi}^H \mathbf{z}' = \mathbf{z}''^H \boldsymbol{\Lambda}\mathbf{z}'' = \sum_{i=1}^{U_{total}} \lambda_i \left|z_i''\right|^2, \tag{4.13}$$

where two more vectors, $\mathbf{z}''^H = \mathbf{z}'^H \boldsymbol{\Phi}$ and $\mathbf{z}'' = \boldsymbol{\Phi}^H \mathbf{z}'$, have been defined, and whose components are still uncorrelated unit-variance complex Gaussian variables. According to (4.13), $\kappa$ can be expressed as a sum of independent random variables with exponential distributions whose characteristic function is

$$E\left\{e^{-jv\kappa}\right\} = \prod_{i=1}^{U_{total}} \frac{1}{1 + j\lambda_i v}. \tag{4.14}$$

If there are $U'$ distinct eigenvalues, each with a multiplicity of $q_i$, $i=1...U'$, the inverse Fourier transform can be applied to (4.14) and obtain the PDF of $\kappa$ as

$$p\left(\kappa\right) = \sum_{i=1}^{U'} \sum_{m=1}^{q_i} \frac{A_{i,m}}{\lambda_i^{q_i}\left(q_i - m\right)!\left(m - 1\right)!}\kappa^{m-1}e^{-\frac{\kappa}{\lambda_i}}, \tag{4.15}$$

with

$$A_{i,m} = \left[\frac{\partial^{q_i - m}}{\partial s^{q_i - m}}\left(\prod_{\substack{j=1 \\ j \neq i}}^{U'} \frac{1}{\left(1 + s\lambda_j\right)^{q_j}}\right)\right]_{s = -\frac{1}{\lambda_i}}. \tag{4.16}$$

The average BER can be obtained as

$$P_{b_{av}} = \int_{-\infty}^{+\infty} P_b(\kappa) p(\kappa) d\kappa = \sum_{i=1}^{U'} \sum_{m=1}^{q_i} \frac{A_{i,m}}{\lambda_i^{q_i-m}(q_i-m)!} \left[\frac{1-\mu_i}{2}\right]^m$$
$$\cdot \sum_{r=0}^{m-1} \binom{m-1+r}{r} \left[\frac{1+\mu_i}{2}\right]^r,$$

where

$$\mu_i = \sqrt{\frac{\frac{E_s}{N_0}\lambda_i}{1+\frac{E_s}{N_0}\lambda_i}}. \tag{4.17}$$

## 4.2 System Characterization

In a conventional OFDM scheme, the time-domain block is $\{s_n; n = 0, 1, \ldots, N-1\} = \text{IDFT}\{S_k; k = 0, 1, \ldots, N-1\}$, with $S_k$ denoting the frequency-domain data symbols to be transmitted, associated with the $k^{th}$ subcarrier, and selected from a given constellation (e.g., a QPSK constellation). On the other hand, for an SC-FDE scheme the time-domain symbols $\{s_n; n = 0, 1, \ldots, N-1\}$, are directly selected from the constellation. For both block transmission schemes a cyclic prefix, with length longer than the overall channel impulse response, is appended leading to the signal $\{s_n^{CP}; n = -N_G, \ldots, N-1\}$, which is transmitted over a time-dispersive channel.

Receivers with $N_{Rx}$ diversity branches are also considered, for both schemes. The signal associated with the $l^{th}$ branch is sampled and the cyclic prefix is removed leading to the time-domain block $\{y_n^{(l)}; n = 0, 1, \ldots, N-1\}$. The corresponding frequency-domain block, obtained after an appropriate size-$N$ DFT operation, is $\{Y_k^{(l)}; k = 0, 1, \ldots, N-1\}$, with $Y_k^{(l)}$ given by (3.27), reproduced below for convenience,

$$Y_k^{(l)} = S_k H_k^{(l)} + N_k^{(l)},$$

with $H_k^{(l)}$ denoting the overall channel frequency response between the transmit antenna and the $l^{th}$ receive antenna for the $k^{th}$ frequency, $S_k$ denoting the frequency-domain of the transmitted block and $N_k^{(l)}$ denoting the corresponding channel noise.

## 4.3 Performance Results

This section presents the performance results concerning the impact of the number of multipath components and the diversity on the performance of OFDM and SC-FDE receivers as well as the correspondent MFB. In both cases blocks with $N = 512$ "useful" data symbols, plus an appropriate cyclic prefix, are considered.

The modulation symbols are selected from a QPSK constellation under a Gray mapping rule, and the channel is characterized by an uniform power delay profile (PDP), with $U = U_1 = ... = U_{N_{Rx}}$ equal-power symbol-spaced multipath components, and uncorrelated Rayleigh fading for all diversity branches. However it is important to point out that a similar behavior was observed for other channels such as exponential PDP. The major difference was in the higher number of multipath components needed to have similar results to those of the uniform PDP. This is due to the fact that the number of relevant multipath components is lower for the exponential PDP, since the last ones have much lower power. For the sake of simplicity, a linear power amplification at the transmitter and perfect synchronization and channel estimation at the receiver was also assumed for the studied cases.

The performance results are expressed as functions of $E_b/N_0$, where $N_0$ is the one-sided power spectral density of the noise and $E_b$ is the energy of the transmitted bits (i.e., the degradation due to the useless power spent on the cyclic prefix is not included).

Since we are trying to approach the MFB performance, we always employ a turbo IB-DFE in the coded case.

### 4.3.1 Performance Results without Channel Coding

Fig. 4.1 shows the typical behavior of the BER performance, for an IB-DFE, without channel coding, and a channel with $U = 16$ separable multipath com-

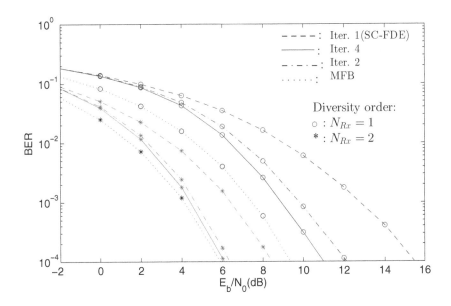

**Figure 4.1: BER performance of an IB-DFE in the uncoded case.**

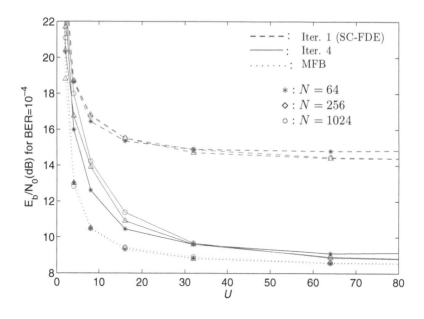

**Figure 4.2:** Required $E_b/N_0$ to achieve $BER = 10^{-4}$ for the uncoded case and without diversity, as a function of the number of multipath components.

ponents for each diversity branch. The SC-FDE employs an IB-DFE receiver with four iterations and the particular case with a single iteration that corresponds to a linear FDE. Clearly, there is a significant performance improvement with the subsequent iterations and the asymptotic BER performance comes closer to the MFB. In this situation the OFDM results are not presented because the uncoded OFDM performance is very poor since OFDM schemes are severely affected by deep-faded subcarriers. In order to analyze the influence of the block size on the performance, an uncoded SC-FDE scheme was considered. As shown in Fig. 4.2, for the case without diversity, the required values of $E_b/N_0$, for a specific BER of $10^{-4}$, are independent of the number of symbols $N$ of each transmitted block.

## 4.3.2 Performance Results with Channel Coding

In what refers to the evaluation of the impact of channel coding, for both modulation schemes a channel encoder was employed, based on a 64-state, 1/2-rate convolutional code with the polynomial generators $1 + D^2 + D^3 + D^5 + D^6$ and $1 + D + D^2 + D^3 + D^6$. Punctured versions of this code (with rates 2/3 and 3/4) are included, with the purpose of increasing the user bit rate while maintaining the gross bit rate [GCG79]. The coded bits are interleaved before

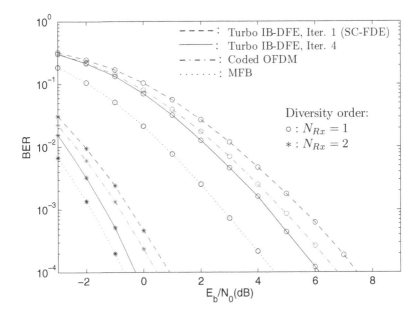

**Figure 4.3: BER performance for a rate-1/2 code.**

being mapped into the constellation points and distributed by the symbols of the block. The first refers to the BER results of a coded transmission considering both an SC-FDE (with a turbo IB-DFE receiver), and OFDM schemes (with the same channel encoder), and a channel with $U = 8$ relevant separable multipath components for each diversity branch. Figs. 4.3, 4.4, and 4.5 consider 1/2-rate, 2/3-rate, and 3/4-rate, respectively. It can be seen from Fig. 4.3 that with a rate-1/2 convolutional code, the OFDM modulation has slightly better performance than SC-FDE with a linear FDE (corresponding to the IB-DFE's first iteration). However, for the following iterations, turbo IB-DFE clearly outperforms OFDM. Figs. 4.4 and 4.5 show the results obtained with the 2/3-rate and 3/4-rate convolutional code, respectively. It is clear that for higher code rates the SC modulation has better performance than OFDM, even with only a single iteration.

Next we present the required values of $E_b/N_0$ for a $BER = 10^{-4}$ for the SC-FDE and OFDM, as well as the corresponding MFB. These values are expressed as a function of the number of multipath components $U$. It can be observed that SC-FDE has an overall performance advantage over the OFDM, especially when employing the IB-DFE with turbo equalization and/or diversity. Therefore, by using SC modulation with turbo equalization, and channels with a high number of multipath components we can be very close to the MFB after a few iterations (naturally, for $U = 1$ the BER is identical to the MFB,

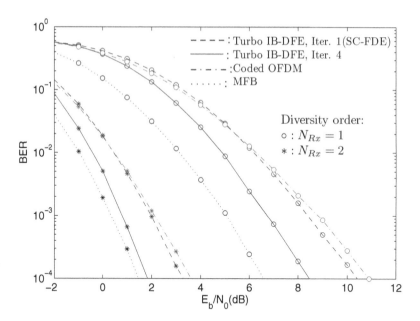

**Figure 4.4: BER performance for a rate-2/3 code.**

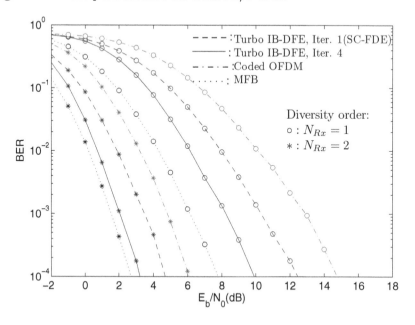

**Figure 4.5: BER performance for a rate-3/4 code.**

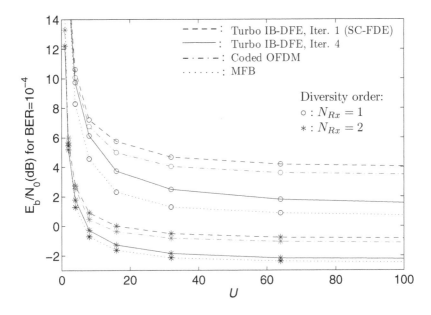

**Figure 4.6: Required $E_b/N_0$ to achieve $BER = 10^{-4}$ for the rate-1/2 convolutional code, as a function of the number of multipath components.**

although the performance is very poor, since this corresponds to a flat fading channel). Observe that the improvements with the iterations are higher without diversity, and this is also the case where a higher number of multipath components is required to allow performances close to the MFB (about $U = 70$).

Finally, we might ask what happens for different fading models? Consider a Nakagami fading with factor $\mu$ on each multipath component (clearly, $\mu = 1$ corresponds to the Rayleigh case and $\mu = +\infty$ corresponds to the case where there is no fading on the different multipath components). Fig. 4.9 presents the required values of $E_b/N_0$ for BER=$10^{-4}$ as a function of the number of multipath components, concerning the MFB and an IB-DFE with 4 iterations.

It should be pointed out that the performance degradation is due to two main factors: the fading effects and the residual ISI. The fading effects in each ray (and, consequently, on the overall received signal) decrease as the Nakagami factor $\mu$ is increased; they also reduce when we increase the number of components due to multipath effects. That is why the MFB is better for larger values of $\mu$ and a larger number of components. The IB-DFE is a very efficient equalizer, being able to reduce significantly the residual ISI effects, especially for a large number of components. Naturally, the overall performance will be the result of these combined effects. For Rayleigh fading channels the fading

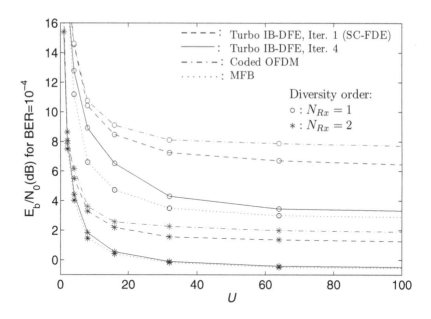

**Figure 4.7: Required $E_b/N_0$ to achieve $BER = 10^{-4}$ for the rate-2/3 convolutional code, as a function of the number of multipath components.**

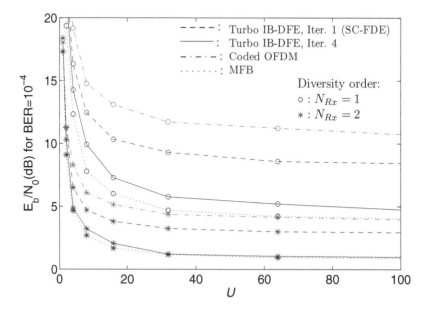

**Figure 4.8: Required $E_b/N_0$ to achieve $BER = 10^{-4}$ for the rate-3/4 convolutional code, as a function of the number of multipath components.**

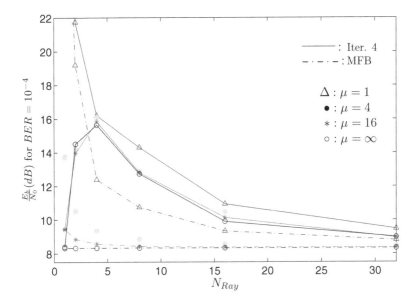

**Figure 4.9:** Required $E_b/N_0$ to achieve $BER = 10^{-4}$ at the MFB and at the $4^{th}$ iteration of the IB-DFE, for an uncoded scenario without diversity and with a Nakagami channel model with factor $\mu$.

effects are very strong and have a higher impact on the performance than the residual ISI effects when we have a small number of multipath components. Therefore, the performance improves steadily as we increase the number of components. When there are smaller fading effects on each ray (as in Nakagami channels with $\mu > 1$), the degradation due to residual ISI becomes more relevant, especially when there is only a small number of components (but more than one). This leads to the somewhat unexpected IB-DFE behavior of Fig. 4.9 where there is a slight degradation as the number of components is increased up to a value, after which there is a steady improvement with the number of components.

With respect to the coded case, the impact of the factor $\mu$ on the performance is negligible for a large number of multipath components. Fig. 4.10 shows the performance of OFDM and SC-FDE (with both a linear FDE and a turbo FDE with 4 iterations), as well as the MFB, for 2, 8, and 32 multipath components. Clearly, the best performance is achieved for the turbo FDE and the worst performance for the linear FDE, with the performance of OFDM schemes somewhere in between. The difference between the MFB and the achieved performance is higher for a moderate number of propagation components (around 8).

It is important to point out that, although obtained by simulation, the results provide important information concerning the achievable performance. It

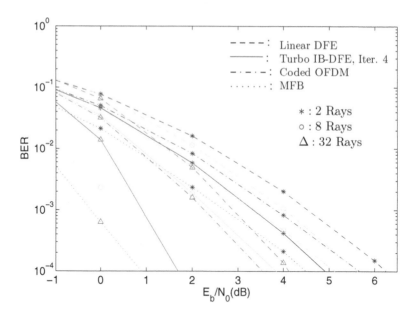

**Figure 4.10: BER performance of OFDM and SC-FDE, for** $U = 2$, **8, and 32, and a Nakagami channel with** $\mu = 4$.

can also be used to decide whether we should employ a more complex IB-DFE or a simple linear FDE: if the number of relevant separable multipath components is very low it is preferable to employ a linear FDE. The above results clearly show that the number of relevant separable multipath components is a fundamental element that influences the performance of both schemes, and in the IB-DFE's case, the gains associated with the iterations. The SC-FDE has an overall performance advantage over OFDM, especially when employing the IB-DFE, and for a high number of separable multipath components, since it allows a performance very close to the MFB, even without diversity. With diversity the performance approaches MFB faster, even for a small number of separable multipath components. In sum, this study shows that the key factor that affects how far the performance of these systems is from the MFB (and in the IB-DFE case, the gains with the iterations) is the number of relevant propagation components.

# Chapter 5

# Efficient Channel Estimation for Single Frequency Networks

Traditional broadcasting systems assign different frequency bands to each transmitter, within a given region in order to prevent interference between transmitters. Frequencies used in a cell will not be allocated in adjacent cells. As an alternative, SFN broadcasting systems [Mat05], where several transmitters transmit the same signal simultaneously and over the same bands, can be employed. Since the distance between a given receiver and each transmitter can be substantially different, the overall channel impulse response can be very long, spanning over hundreds or even thousands of symbols in the case of broadband broadcasting systems; this can cause severe time-distortion effects within this type of single frequency system.

To deal with the severe distortion inherent to SFN, digital broadcasting standards such as digital video broadcasting (DVB) [Rei95] and digital audio broadcasting (DAB) [MR93] use OFDM modulations which are known to be suitable for severely time-dispersive channels.

In SFN broadcasting systems the equivalent CIR can be very long, typically with a sparse nature. This means that the equivalent CIR has several clusters of paths, each one associated with the CIR between a given transmitter and the receiver. These clusters have several multipath components and are typically well separated in time. This chapter considers OFDM-based broadcasting systems with SFN operation and proposes an efficient channel estimation method that takes advantage of the sparse nature of the equivalent

CIR. For this purpose, low-power training sequences are employed within an iterative receiver which performs joint detection and channel estimation.

The receiver operation is based on the assumption that the receiver can know the location of the different clusters that constitute the overall CIR. Nevertheless, several methods were proposed for the case where the receiver does not know the location of the clusters that constitute the overall CIR.

# 5.1 System Characterization

In conventional broadcasting systems each transmitter serves a cell and the frequencies used in a cell are not used in adjacent cells. Typically, this means a frequency reuse factor of 3 or more [CKB06], leading to an inefficient spectrum management since the overall bandwidth required for the system is the required bandwidth for a given transmitter times the reuse factor. However, the system's spectral efficiency can be improved significantly if multiple transmitters employ the same frequency. In an SFN scenario [Mat05], the transmitters transmit simultaneously the same signal on the same frequency band, allowing a high spectral efficiency, leading to a reuse factor of 1 (see Fig. 5.1).

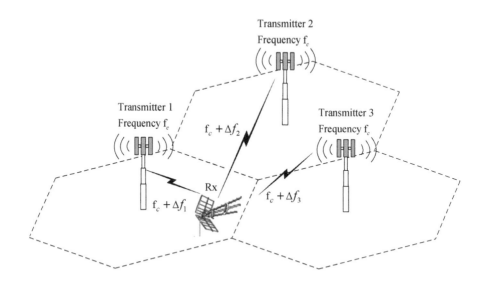

**Figure 5.1: Single frequency network transmission.**

However, the SFN transmission causes time dispersion mainly induced by two factors: the natural multipath propagation due to the reflected or refracted waves in the neighborhood of the receiver, and the unnatural multipath propagation effect due to the reception of the same signals from multiple transmitters, which are added being the resulting signal equivalent to consider a transmission over a single time-dispersive channel. These signals can be seen as "artificial echoes." The receiver's performance can be compromised, since the frequency selective fading may cause very low values of the instantaneous SNR at the receiver.

As referred to before, OFDM has been used as the modulation technique in SFN in order to prevent multipath propagation. The data rate in DVB systems is very high, which means that the overall channel impulse response can span over hundreds or even thousands of symbols. This means that we need to employ very large fast Fourier transform blocks (FFT) to avoid significant degradation due to the cyclic prefix. The DVB standard considers up to 8k-length blocks, corresponding to several thousands of subcarriers. Coherent receivers are usually assumed in a broadcasting system, which means that accurate channel estimates are required at the receiver. The channel can be estimated with the help of pilots or training blocks [SDM10]. The frequency selective fading can be mitigated by employing equalization and/or coding techniques.

Assume the frame structure depicted in Fig. 5.2, with a training block followed by $N_D$ data blocks, each one corresponding to an "FFT block," with $N$ subcarriers. Both the training and the data blocks are preceded by a cyclic prefix whose duration $T_{CP}$ is longer than the duration of the overall channel impulse response (including the channel effects and the transmit and receive filters). The duration of the data blocks is $T_D$, each one corresponding to a size-$N$ DFT block, and the duration of the training blocks is $T_{TS}$, which can be equal to or smaller than $T_D$. To simplify the implementation we will assume that $T_{TS} = T_D/L$ where $L$ is a power of 2, which means that the training sequence will be formally equivalent to having one pilot for each $L$ subcarriers when the channel is static. The overall frame duration is $T_F = (N_D+1)T_{CP} + T_{TS} + N_D T_D$. If the channel is almost invariant within the frame, the training block can provide the channel frequency response for the subsequent $N_D$ data blocks. When it can be afforded a delay of about half the frame duration then it becomes possible to use the training block to estimate the channel for the $N_D/2$ blocks before and after the training, grossly duplicating the robustness to channel variations.[1]

---

[1]For fast-varying channels, it is required to interpolate channel estimates resulting from different training sequences, although increasing significantly the delay (in this case delays of several frames may be needed). With an ideal $sinc()$ interpolation the maximum Doppler frequency is around $1/(2T_F)$.

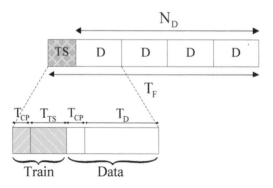

**Figure 5.2: Frame structure.**

The transmitted signal associated with the frame is

$$s^{Tx}(t) = \sum_{m=1}^{N_D} s^{(m)}(t - mT_B),\qquad(5.1)$$

with $T_B$ denoting the duration of each block. The $m^{th}$ transmitted block has the form

$$s^{(m)}(t) = \sum_{n=-N_G}^{N-1} s_n^{(m)} h_T(t - nT_S),\qquad(5.2)$$

with $T_S$ denoting the symbol duration, $N_G$ denoting the number of samples at the cyclic prefix, and $h_T(t)$ is the adopted pulse shaping filter. Clearly, $T_S = T_D/N$ and $N_G = T_{CP}/T_S$.

In a conventional OFDM scheme, the $m^{th}$ time-domain block is $\{s_n^{(m)}; n = 0, 1, \ldots, N-1\} = \text{IDFT } \{S_k^{(m)}; k = 0, 1, \ldots, N-1\}$, with $S_k^{(m)}$ denoting the frequency-domain data symbols to be transmitted, selected from a given constellation (e.g., a QPSK constellation), and associated with the $k^{th}$ subcarrier. The signal $s^{(m)}(t)$ is transmitted over a time-dispersive channel, leading to the time-domain block $\{y_n^{(m)}; n = 0, 1, \ldots, N-1\}$, after cyclic prefix removal. The corresponding frequency-domain block, obtained after an appropriate size-$N$ DFT operation, is $\{Y_k^{(m)}; k = 0, 1, \ldots, N-1\}$, where

$$Y_k^{(m)} = S_k^{(m)} H_k^{(m)} + N_k^{(m)},\qquad(5.3)$$

with $H_k^{(m)}$ denoting the overall channel frequency response for the $k^{th}$ frequency of the $m^{th}$ time block and $N_k^{(m)}$ denoting the corresponding channel noise. Clearly, the impact of the time dispersive channel reduces to a scaling factor for each frequency. For the sake of simplicity, slow-varying channel will be assumed, i.e., $H_k^{(m)} = H_k$.

## 5.1.1 Channel Estimation

Since the optimum FDE coefficients are a function of the channel frequency response, accurate channel estimates are required at the receiver. To improve the channel estimation performance a joint detection and channel estimation [LMWA02, CH03] can be done. To avoid performance degradation the power spent in training blocks should be similar to or higher than the power associated with the data. However, there is always some performance degradation when the power spent to transmit each block, i.e., the power of training plus data, is considered.

As with data blocks, the training signal has the form

$$s^{TS}(t) = \sum_{n=-N_{CP}}^{N_{TS}-1} s_n^{TS} h_T(t - nT_S), \tag{5.4}$$

where $s_n^{TS}$ denotes the $n^{th}$ symbol of the training sequence, and the corresponding time-domain block at the receiver, after cyclic prefix removal, will be $\{y_n^{TS}; n = 0, 1, \ldots, N_{TS} - 1\}$. The corresponding frequency-domain block $\{Y_k^{TS}; k = 0, 1, \ldots, N_{TS} - 1\}$ is the size-$N_{TS}$ DFT of $\{y_n^{TS}; n = 0, 1, \ldots, N_{TS} - 1\}$. Since $N_{TS} = N/L$, it can be written

$$Y_k^{TS} = S_k^{TS} H_{kL} + N_k^{TS}, \ k = 0, 1, ..., N_{TS} - 1, \tag{5.5}$$

with $\{S_k^{TS}; k = 0, 1, \ldots, N_{TS} - 1\}$ denoting the size-$N_{TS}$ DFT of $\{s_n^{TS}; n = 0, 1, \ldots, N_{TS} - 1\}$ and $N_k^{TS}$ denoting the channel noise. The channel frequency response could be estimated as follows:

$$\tilde{H}_{kL} = \frac{Y_k^{TS}}{S_k^{TS}} = H_{kL} + \frac{N_k^{TS}}{S_k^{TS}} = H_{kL} + \epsilon_{kL}^H, \tag{5.6}$$

where the channel estimation error, $\epsilon_{kL}^H$ is Gaussian-distributed, with zero-mean.

It should be noted that, when $L > 1$, it will be necessary to interpolate the channel estimates. In this case, it is necessary to form the block $\{\tilde{H}_k^{TS}; k = 0, 1, \ldots, N - 1\}$, where $\tilde{H}_k^{TS} = 0$ when $k$ is not a multiple of $L$ (i.e., for the subcarriers that do not have estimates given by (5.6)) and compute its IDFT, to derive $\{\tilde{h}_n^{TS}; n = 0, 1, \ldots, N - 1\}$. Provided that the channel impulse response is restricted to the first $N_{CP}$ samples, the interpolated channel frequency response is $\{\hat{H}_k^{TS}; k = 0, 1, \ldots, N - 1\} = \text{DFT}$ $\{\hat{h}_n^{TS} = \tilde{h}_n^{TS} w_n; n = 0, 1, \ldots, N - 1\}$, where $w_n = 1$ if the $n^{th}$ time-domain sample is inside the cyclic prefix (first $N_{CP}$ samples) and 0 otherwise. Naturally,

$$\hat{H}_k^{TS} = H_k + \epsilon_k^{TS}, \tag{5.7}$$

where $\epsilon_k^{TS}$ represents the channel estimation error after the interpolation. It can be shown that $\epsilon_k^{TS}$ is Gaussian-distributed, with zero-mean and $E[|\epsilon_k^{TS}|^2] = \sigma_{H,TS}^2 = \sigma_N^2 |S_k^{TS}|^2$, assuming $|S_k^{TS}|$ constant. Since the power assigned to the training block is proportional to $E[|S_k^{TS}|^2] = \sigma_T^2$ and $E\left[1/|S_k^{TS}|^2\right] \geq 1/E[|S_k^{TS}|^2]$, with equality for $|S_k^{TS}|$ constant, the training blocks should have $|S_k^{TS}|^2 = \sigma_T^2$ for all $k$. By contrast, if it is intended to minimize the envelope fluctuations of the transmitted signal the value of $|s_n^{TS}|$ should be also constant. This condition can be achieved by employing Chu sequences, which have both $|s_{n,m}^{TS}|$ and $|S_{k,m}^{TS}|$ constant [Chu72].

Since the channel impulsive response is usually shorter than the cyclic prefix, training blocks shorter than the data blocks, could be employed. As an alternative, a training block with the same duration of the data block $(N = N_{TS})$ can be used, which is typically much longer than duration of the channel impulse response, and employ the enhanced $\{\hat{H}_k^{TS}; k = 0, 1, \ldots, N-1\}$ $= \text{DFT} \{\hat{h}_n^{TS} = \tilde{h}_n^{TS} w_n; n = 0, 1, \ldots, N - 1\}$, with $w_n$ defined as above and $\{\tilde{h}_n^{TS}; n = 0, 1, \ldots, N - 1\} = \text{IDFT} \{\tilde{H}_k^{TS} = Y_k^{TS}/S_k^{TS}; k = 0, 1, \ldots, N - 1\}$. In this case, the noise's variance in the channel estimates, $\sigma_{H,TS}^2$, is improved by a factor $N/N_{CP}$. Naturally, the system's spectral efficiency decreases (due to the use of longer training sequences) and the overall power spent in the training sequence increases, although the power per subcarrier and the peak power remain the same.

### 5.1.2 Channel Estimation Enhancement

As stated above, the SFN transmission creates severe artificial multipath propagation conditions. Typically the SFN systems employ a large number of OFDM subcarriers, to ensure that the guard interval is large enough to cope with the maximum delay spread that can be handled by receivers. In fact this measure partly determines how far apart transmitters can be placed in the SFN. Although the system is defined to accommodate the worst case scenario (i.e., maximum delay spread), it also may represent a waste of bandwidth in most cases.

In this section several methods to improve spectral efficiency in the channel estimation given by (5.6) are proposed. To better understand the involved operations all methods are associated with a figure that illustrates the impact of the enhancement process on the channel's impulsive response. An

OFDM modulation with blocks of $N = 8192$ "useful" modulation symbols is considered plus a cyclic prefix of 2048 symbols acquired from each block (corresponding to the OFDM 8K mode in DVB-T). Also considered is a channel's impulsive response corresponding to a sum of three identical signals emitted from three transmitters and received with different delays and power.

## *Method I*

The first method employs the basic filtering operation given by the enhanced channel frequency response $\{\hat{H}_k^{TS}; k = 0, 1, \ldots, N - 1\} = \text{DFT}$ $\{\hat{h}_n^{TS} = \tilde{h}_n^{TS} w_n; n = 0, 1, \ldots, N - 1\}$, where $w_n = 1$ if the $n^{th}$ time-domain sample is inside the cyclic prefix (first $N_{CP}$ samples) and 0 otherwise. The overall CIR is depicted in Fig. 5.3. Considering $N$ useful modulation symbols and a cyclic prefix of $N_{CP}$ symbols, the resulting gain associated with this method is $G_1 = N/N_{CP}$.

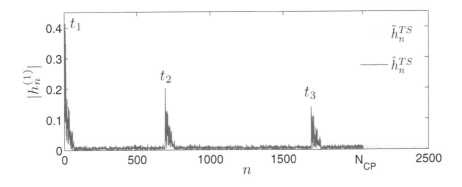

**Figure 5.3: Impulsive response of the channel estimation with method I.**

## *Method II*

This method is specific for SFN, and it assumes that the CIR related to each one of the three channels is perfectly known (i.e., the receiver knows the exact duration $\Delta N$ of each CIR, and the location of the different clusters). The overall CIR is depicted in Fig. 5.4. Since $\Delta N >> N_{CP}$, the gain associated with this method, given by $G_2 = N/\Delta N$, is much higher than the gain of method I, (i.e., $G_1 >> G_2$).

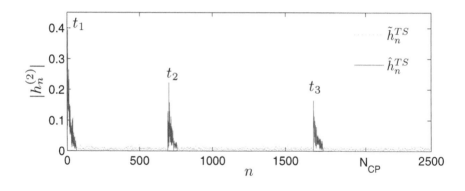

**Figure 5.4: Impulsive response of the channel estimation with method II.**

## Method III

This method assumes that whenever a sample corresponds to a relevant multipath component (i.e., a strong ray), a small set of samples before and after that sample must be considered. The overall CIR is depicted in Fig. 5.5.

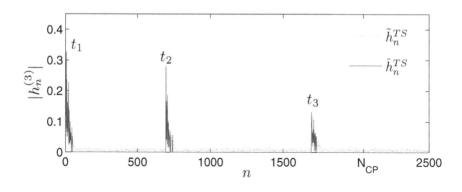

**Figure 5.5: Impulsive response of the channel estimation with method III.**

## Method IV

This method considers as relevant only the multipath components whose power exceeds a pre-defined threshold. The samples below this limit are considered as noise and ignored. The overall CIR is depicted in Fig. 5.6.

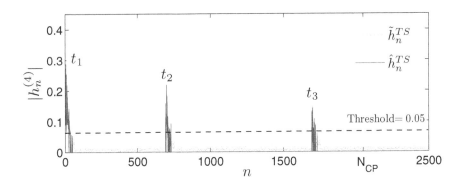

**Figure 5.6: Impulsive response of the channel estimation with method IV.**

## 5.2 Decision-Directed Channel Estimation

The channel estimation methods described above are based on training sequences multiplexed with data. To avoid performance degradation due to channel estimation errors the required average power for these sequences should be several dB above the data power.[2] Here it is shown how it is possible to use a decision-directed channel estimation to improve the accuracy of channel estimates without resort to high-power training sequences.

If the transmitted symbols for a set of $N_D$ data blocks $\{S_k^{(m)}; k = 0, 1, ..., N - 1\}$ $(m = 1, 2, ..., N_D)$ were known in advance, the channel could be estimated as follows

$$\tilde{H}_k^D = \frac{\sum_{m=1}^{N_D} Y_k^{(m)} S_k^{(m)*}}{\sum_{m=1}^{N_D} |S_k^{(m)}|^2} = H_k + \frac{\sum_{m=1}^{N_D} N_k^{(m)} S_k^{(m)*}}{\sum_{m=1}^{N_D} |S_k^{(m)}|^2}. \tag{5.8}$$

This basic channel estimate $\{\tilde{H}_k^D; k = 0, 1, ..., N - 1\}$ can be enhanced as described for the case where $N_{TS} = N$: from $\{\tilde{h}_n^D; n = 0, 1, ..., N - 1\}$ = IDFT $\{\tilde{H}_k^D; k = 0, 1, ..., N - 1\}$ is obtained $\{\hat{H}_k^D; k = 0, 1, ..., N - 1\}$ = DFT $\{\hat{h}_n^D = \tilde{h}_n^D w_n; n = 0, 1, ..., N - 1\}$, with $w_n$ defined as above. Henceforward, the term "enhanced channel estimates" will be adopted to characterize this procedure (starting with estimates for all subcarriers, passing to the time domain where the impulse response is truncated to $N_{CP}$ samples and back to the frequency domain). Clearly,

$$\hat{H}_k^D = H_k + \epsilon_k^D, \tag{5.9}$$

---

[2] As mentioned before, the use of training blocks longer than the channel impulse response (e.g., with the duration of data blocks) can improve the accuracy of the channel estimates, but it reduces the system's spectral efficiency.

with

$$E[|\epsilon_k^D|^2] = \sigma_D^2 = \frac{N_{CP}\sigma_N^2}{N\sum_{m=1}^{N_D}|S_k^{(m)}|^2}. \tag{5.10}$$

The channel estimates obtained from the training sequence are $\tilde{H}_k^{TS} = H_k + \epsilon_k^{TS}$, with variance $\sigma_{TS}^2 = \sigma_N^2/|S_k^{TS}|^2$ (for the sake of simplicity, it is assumed that the duration of the training sequences is equal to the duration of the channel impulse response, i.e., $T_{CP} = T_D/L$, with $L$ a power of 2). As described in Appendix C, $\tilde{H}_k^{TS}$ and $\tilde{H}_k^D$ can be combined to provide the normalized channel estimates with minimum error variance, given by

$$\tilde{H}_k^{TS,D} = \frac{\sigma_D^2\tilde{H}_k^{TS} + \sigma_{TS}^2\tilde{H}_k^D}{\sigma_D^2 + \sigma_{TS}^2} = H_k + \epsilon_k^{TS,D}, \tag{5.11}$$

with

$$E[|\epsilon_k^{TS,D}|^2] = \sigma_{TS,D}^2 = \frac{\sigma_D^2\sigma_{TS}^2}{\sigma_D^2 + \sigma_{TS}^2}. \tag{5.12}$$

Naturally, in realistic conditions the transmitted symbols are not known. To overcome this problem, a decision-directed channel estimation can be employed, where the estimated blocks $\{\hat{S}_k^{(m)}; k = 0, 1, , ..., N - 1\}$ are used in place of the transmitted blocks $\{S_k^{(m)}; k = 0, 1, , ..., N - 1\}$. Moreover, it must be taken into account the fact that there can be decisions errors in the data estimates. This can be done by noting that $\hat{S}_k^{(m)} \approx \rho_m S_k^{(m)} + \Delta_k^{(m)}$, where $\rho_m$ refers to the correlation coefficient of the $m^{th}$ data block, and $\Delta_k^{(m)}$ the zero-mean error term for the $k^{th}$ frequency-domain "hard decision" estimate of the $m^{th}$ data block. Note that $\Delta_k^{(m)}$ is uncorrelated with $S_k^{(m)}$ and $E[|\Delta_k^{(m)}|^2] = \sigma_S^2(1 - \rho_m^2)$ [DGE03], meaning that the "enhanced channel estimates" $\hat{H}_k^D$ will be based on

$$\tilde{H}_k^D = \frac{1}{\xi_k}\sum_{m=1}^{N_D} Y_k^{(m)}\hat{S}_k^{(m)*}, \tag{5.13}$$

with

$$\xi_k = \sum_{m=1}^{N_D} |\rho_m\hat{S}_k^{(m)}|^2. \tag{5.14}$$

Replacing $\hat{S}_k^{(m)}$ and $Y_k^{(m)}$ in (5.13) results in

$$\begin{aligned}
\tilde{H}_k^D &= \frac{1}{\xi_k}\sum_{m=1}^{N_D}(S_k^{(m)}H_k + N_k^{(m)})(\rho_m S_k^{(m)} + \Delta_k^{(m)})^* \\
&= \frac{H_k}{\xi_k}\sum_{m=1}^{N_D}\rho_m|S_k^{(m)}|^2 + \frac{1}{\xi_k}(H_k\sum_{m=1}^{N_D}S_k^{(m)}\Delta_k^{(m)*} \\
&\quad + \sum_{m=1}^{N_D}N_k^{(m)}\rho_m S_k^{(m)*} + \sum_{m=1}^{N_D}N_k^{(m)}\Delta_k^{(m)*}).
\end{aligned} \tag{5.15}$$

It can easily be shown that $\hat{H}_k^D = H_k + \epsilon_k^D$, with

$$E[|\epsilon_k^D|^2] = \sigma_D^2 = \frac{1}{\xi_k^2}(|H_k|^2 \sum_{m=1}^{N_D} |S_k^{(m)}|^2 (1 - \rho_m^2)\sigma_S^2 +$$

$$\sum_{m=1}^{N_D} \sigma_N^2 \rho_m^2 |S_k^{(m)}|^2 + \sum_{m=1}^{N_D} \sigma_N^2 (1 - \rho_m^2)\sigma_S^2)$$

$$\approx \frac{1}{\xi_k^2}(|\hat{H}_k|^2 \sum_{m=1}^{N_D} |\hat{S}_k^{(m)}|^2 (1 - \rho_m^2)\sigma_S^2 +$$

$$\sum_{m=1}^{N_D} \sigma_N^2 \rho_m^2 |\hat{S}_k^{(m)}|^2 + \sum_{m=1}^{N_D} \sigma_N^2 (1 - \rho_m^2)\sigma_S^2)$$

(5.16)

## 5.3 Performance Results

In this section, a set of performance results concerning the proposed channel estimation method for single frequency broadcast systems is presented and analyzed. It is assumed that the identical signals emitted from three different transmitters arrive at the receiver with different delays. At the receiver antenna, these signals are added, being the resulting signal equivalent to the result of a transmission over a single strong time-dispersive channel. The typical delay profile for this channel is similar to the one presented in Figs. 5.3, 5.4, and 5.5.

An OFDM modulation is considered, with blocks of $N = 8192$ subcarriers and a cyclic prefix of 2048 symbols acquired from each block. The modulation symbols belong to a QPSK constellation (on each subcarrier) and are selected from the transmitted data according to a Gray mapping rule. Similar results were observed for other values of $N$, provided that $N \gg 1$.

A coded transmission employing a channel encoder based on a 64-state, 1/2-rate convolutional code with the polynomial generators $1 + D^2 + D^3 + D^5 + D^6$ and $1 + D + D^2 + D^3 + D^6$ was considered. The coded bits were interleaved before being mapped into the constellation points and distributed by the symbols of the block. Linear power amplification at the transmitter and perfect synchronization at the receiver were also assumed. The performance results are expressed as functions of $E_b/N_0$, where $N_0$ is the one-sided power spectral density of the noise and $E_b$ is the energy of the transmitted bits.

The following figures present a set of performance results for the proposed channel estimation technique based on the enhancement methods discussed in Sec. 5.1.2. For comparison purposes, the BER performance results for perfect channel estimation, were also included. The impact of the relation between the average power of the training sequences, and the data power, is also evaluated. The relation is denoted by $\beta$, in the asymptotic performance.

Figs. 5.7 and 5.8 present the BER results for $N_D = 1$ and $N_D = 4$, respectively, with $\beta = 1/16$.

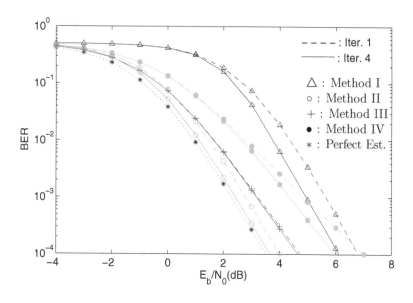

**Figure 5.7: BER performance for OFDM with $N_D = 1$ block and $\beta = 1/16$.**

Clearly, for both cases, the best performance results can be achieved when method II is adopted to improve the channel estimate obtained from the training sequence. This method assumes that the receiver knows the location of the different clusters that constitute the overall CIR. In fact, the results are very close to those with perfect estimation. Also, the impact of the iteration number in power efficiency is higher for longer frame structures, i.e., higher number of data blocks. This effect is clearly seen in the higher power gains achieved by the iterative process when a frame with $N_D = 4$ blocks is used. Obviously, this is due to the fact that the channel estimates are more accurate for larger frames, i.e., when more data blocks are used in the decision-directed estimation. This is a consequence of the higher power of the overall signals, as well as the lower probability of $\sum_{m=1}^{N_D} |S_k^{(m)}|^2 \approx 0$ when $N_D$ is high.

Lastly, regarding the impact of $\beta$ in the asymptotic performance, results in terms of the useful $E_b/N_0$ are presented that include only the power spent on the data and denoted as $E_U$ as well as the results in terms of the total $E_b/N_0$, denoted as $E_{Tot}$, which include the degradation associated with the power spent on the training sequence and the power spent on the cyclic prefix, for both the training and the data. For comparison purposes, in Fig. 5.9 and Fig.

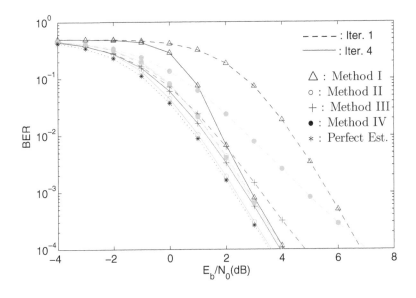

**Figure 5.8: BER performance for OFDM with $N_D = 4$ block and $\beta = 1/16$.**

5.10 are shown the $E_U$ and $E^{Tot}$ needed to assure a BER=$10^{-4}$, for $N_D$=1 and for the $4^{th}$ iteration. From these figures, it can be concluded that the methods II, III, and IV are very robust since they demonstrate performance results almost independent of $\beta$. Therefore these methods allow good initial channel estimates even when employing very low-power training sequences.

## 5.4 Conclusions

The results considered channel estimation for OFDM-based broadcasting systems with SFN operation and we proposed an efficient channel estimation method that takes advantage of the sparse nature of the equivalent CIR. For this purpose, we employed low-power training sequences to obtain an initial coarse channel estimate and we employed an iterative receiver with joint detection and channel estimation. It was also assumed that the receiver can know the location of the different clusters that constitute the overall CIR or not. The performance results show that very good performance, close to the performance with perfect channel estimation, can be achieved with the proposed methods, even when low-power training blocks are employed and the receiver does not know the location of the different clusters that constitute the overall CIR.

We have seen that efficient and accurate channel estimation is mandatory

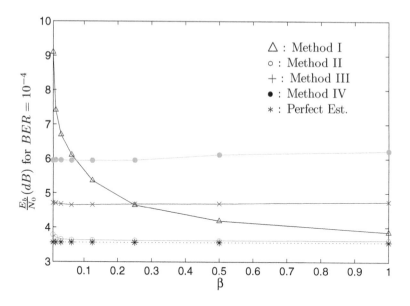

**Figure 5.9:** Useful $E_b/N_0$ required to achieve $BER = 10^{-4}$ with $N_D = 1$, as function of $\beta$: **OFDM for the** $4^{th}$ **iteration.**

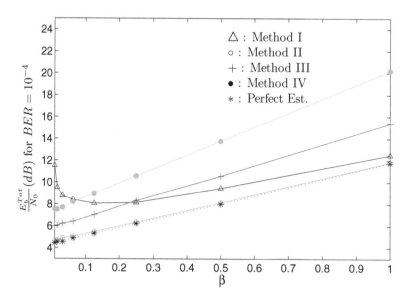

**Figure 5.10:** Total $E_b/N_0$ required to achieve $BER = 10^{-4}$ with $N_D = 1$, as function of $\beta$: **OFDM for the** $4^{th}$ **iteration.**

for the good performance of FDE receivers, both for OFDM and SC-FDE. However, when the channel changes within the block duration then significant performance degradation may occur. The channel variations lead to two different difficulties: first, the receiver needs continuously accurate channel estimates; second, conventional receiver designs for block transmission techniques are not suitable when there are channel variations within a given block. It is therefore difficult to ensure stationarity of the channel within the block duration, which is a requirement for conventional OFDM and SC-FDE receivers.

The following chapters will propose efficient estimation and tracking procedures which will be able to cope with channel variations.

Chapter 5, in part, is a reprint of the paper "Efficient Channel Estimation for Single Frequency Broadcast Systems," F. Silva, R. Dinis, and P. Montezuma, published in the *Vehicular Technology Conference (VTC Fall), 2011 IEEE*, vol. 1, no. 6, pp. 5-8, Sept. 2011.

# Chapter 6

# Asynchronous Single Frequency Networks

To cope with the severe time-distortion effects inherent to SFN systems, most conventional broadband broadcast and multicast wireless systems employing digital broadcasting standards selected OFDM schemes [Cim85], which are known to be suitable to severely time-dispersive channels.

However, OFDM signals have large envelope fluctuations and high PAPR leading to amplification difficulties [MAG98, DG04]. Moreover, due to the very small subcarrier spacing, which is a small fraction of the transmission bandwidth, the carrier synchronization demands in OFDM modulations are very high. A small carrier frequency offset compromises the orthogonality between the OFDM subcarriers, leading to performance degradation that increases rapidly with the frequency offset. An alternative approach based on the same block transmission principle is SC-FDE. As stated before, SC-FDE signals have the advantage of reduced envelope fluctuations due to the much lower envelope fluctuations than OFDM signals based on the same constellation, allowing efficient and low-complexity transmitter implementations [GDCE00,FABSE02]. When compared with OFDM, SC-FDE has the advantage of reduced envelope fluctuations and higher robustness to carrier frequency errors (contrary to OFDM schemes, where frequency errors lead to ICI [6], for SC-FDE the CFO induces a rotation in the constellation that grows linearly along the block). The performance of SC-FDE can be improved with resort to the IB-DFE, and it was shown in Chapter 4 that under certain circumstances, it provides performances close to the MFB in severely time-dispersive channels. For these reasons SC-FDE schemes have been recently proposed for several broadband wireless systems [WYWS10, PDN10, DMCG12].

OFDM and SC-FDE transmit data in blocks and a suitable CP, longer than the maximum expected overall CIR length is appended to each block.

However, due to the very long overall channel impulse response in broadband wireless broadcasting systems, very large blocks with hundreds or even thousands of symbols, are needed. In these conditions, it becomes difficult to ensure that the channel is stationary within the block duration, a requirement for conventional OFDM and SC-FDE receivers. If the channel changes within the block duration we can have significant performance degradation. The channel variations can be a consequence of two main factors, the Doppler effects associated with the relative motion between the transmitter and the receiver [JCWY10] and/or the frequency errors between the local oscillators at the transmitter and the receiver, due to phase noise or residual CFO. Oscillator drifts consist of frequency errors due to frequency mismatch between the local oscillators at the transmitter and receiver. This affects the coherent detection of the transmitted signal by inducing a phase rotation on the equivalent channel that changes within the block, which is equivalent to saying that it varies with time. And that is the reason why the channel affected by CFO is said to vary in time. Obviously, unless dealt with, these channel variations lead to performance degradation regardless of the block transmission technique [DLF04]. Nevertheless, while this residual CFO leads to simple phase variations that can be easily estimated and canceled at the receiver, with resort to the conventional techniques [SF08, AD04, DAPN10], Doppler effects are harder to treat. However, for typical systems the maximum Doppler offset is much lower than the CFO, which means a lower impact on the performance. But that is not the case for SFN systems, where simultaneous transmitters may have different CFOs, which leads to a very difficult scenario where substantial variations on the equivalent channel may happen due to phase variations that cannot be treated as simple phase rotations. Even when the channel is assumed as static, there can be carrier synchronization issues between the different transmitters due to the existence of frequency mismatches between the local oscillator at each transmitter and the local oscillator at the receiver.

Several techniques were proposed for estimating the residual CFO in OFDM schemes [Moo94, SC97, MM99]. In [Moo94], a maximum likelihood frequency offset estimation technique was proposed. This method is based on the repetition of two similar symbols, with a frequency acquisition range $\pm 1/(2T)$, where $T$ is the "useful" symbol duration. An estimator based on the best linear unbiased estimator (BLUE) principle, and requiring one training symbol with $L > 2$ similar parts, and with a frequency acquisition range $\pm L/(2T)$, was proposed in [MM99].

# 6.1 SFN Channel Characterization

Focusing on the transmission of a signal $s(t)$ through the SFN system, consider the ideal case in which different transmitters emit exactly the same signal without CFO (i.e., assuming a perfect carrier synchronization between all transmitters).

The channel's impulse response corresponding to the $l^{th}$ transmitter is given by

$$h^{(l)}(t) = \sum_{i=1}^{N_{Ray}} \alpha_i^{(l)} \delta\left(t - \tau_i^{(l)}\right), \tag{6.1}$$

where $\alpha_i^{(l)}$ and $\tau_i^{(l)}$ are the complex gain and delay associated with the $i^{th}$ multipath component of the $l^{th}$ transmitter (without loss of generality it is assumed that all channels have the same numbers of multipath components).

The equivalent channel's impulse response at the receiver side can be seen as the sum of the impulse responses corresponding to the $N_{Tx}$ transmitters, and can be defined as

$$h(t) = \sum_{l=1}^{N_{Tx}} h^{(l)}(t), \tag{6.2}$$

while the received signal waveform $y(t)$ is the convolution of $s(t)$ with the equivalent channel's impulse response, $h(t)$, plus the noise signal $\nu(t)$, i.e.,

$$\begin{aligned} y(t) = s(t) * h(t) + \nu(t) &= \sum_{l=1}^{N_{Tx}} s(t) * h^{(l)}(t) + \nu(t) \\ &= \sum_{l=1}^{N_{Tx}} y^{(l)}(t) + \nu(t), \end{aligned} \tag{6.3}$$

with $\nu_l$ representing AWGN samples with unilateral power spectral density $N_0$. The signal $y(t)$ is sampled at the receiver, and the CP is removed, leading to the time-domain block $\{y_n; n = 0, ..., N-1\}$, with

$$y_n = \sum_{l=1}^{N_{Tx}} y_n^{(l)} + \nu_n. \tag{6.4}$$

Since the corresponding frequency-domain block associated with the $l^{th}$ transmitter, obtained after an appropriate size-$N$ DFT operation, is $\{Y_k^{(l)}; k = 0, 1, \ldots, N-1\}=$DFT$\{y_n^{(l)}; n = 0, 1, \ldots, N-1\}$, we may write

$$Y_k = \sum_{l=1}^{N_{Tx}} Y_k^{(l)} + N_k = S_k H_k + N_k, \tag{6.5}$$

where

$$H_k = \sum_{l=1}^{N_{Tx}} H_k^{(l)}, \tag{6.6}$$

with $H_k^{(l)}$ denoting the channel frequency response for the $k^{th}$ subcarrier of the $l^{th}$ transmitter.

## 6.2 Impact of Carrier Frequency Offset Effects

The adoption of SFN architectures leads to additional implementation difficulties, mainly due to the synchronization requirements. The fact that the equivalent channel is the sum of the channels associated with each transmitter, with substantially different delays and each one with different multipath propagation effects [Mat05], it is also required to cope with severely time-dispersive channels. This section is dedicated to the analysis of the impact of different CFO between the local oscillator at each transmitter and the local oscillator at the receiver. For the sake of simplicity, it is assumed that each transmission is affected by a corresponding CFO that induces a phase rotation which grows linearly along the block [PDN10]. Without loss of generality, it is assumed that, for each transmitter, the phase rotation is 0 for the initial sample ($n = 0$).[1] In this case, the received equivalent time-domain block, consists of the addition of the time-domain blocks associated with the $N_{Tx}$ transmitters, and is given by $\{y_n^{(\Delta f)}; n = 0, 1, \ldots, N-1\} = \text{IDFT}\{Y_k^{(\Delta f)}; k = 0, 1, \ldots, N-1\}$, where

$$Y_k^{(\Delta f)} = \sum_{l=1}^{N_{Tx}} Y_k^{(\Delta f^{(l)})} + N_k = \sum_{l=1}^{N_{Tx}} S_k^{(\Delta f^{(l)})} H_k^{(l)} + N_k, \quad (6.7)$$

with the block $\{S_k^{(\Delta f^{(l)})}; k = 0, 1, \ldots, N-1\}$ denoting the DFT of the block $\{s_n^{(\Delta f^{(l)})} = s_n e^{j\theta_n^{(l)}}; n = 0, 1, \ldots, N-1\}$, i.e., the original data block with the appropriate phase rotations. The equivalent transmission model is presented in Fig. 6.1. where $\theta_n^{(l)}$ denotes the phase rotation associated with the $l^{th}$ transmitter, $\Delta f^{(l)}$ represents the CFO for the $l^{th}$ transmitter and $\nu(t)$ the noise signal.

## 6.3 Channel and CFO Estimation

Frequency errors in OFDM schemes lead to ICI [WW05a], and in order to mitigate this problem, two estimation techniques were proposed in [WW05a] and [WW05b]. An efficient equalization technique was also proposed in [WHW08].

The impact of CFO errors is serious in SFN broadcasting systems because there can be a different CFO between the local oscillator at each transmitter and the local oscillator at the receiver, which means that even in static channels we can have variations on the equivalent channel frequency response that are not simple phase rotations (which can be easily estimated and canceled at the receiver), and for this reason conventional CFO estimation techniques such as the ones of [Moo94, SC97, MM99] are not appropriate for estimating the different CFOs inherent to SFN scenarios.

---

[1] Clearly, the initial phase rotation can be absorbed by the channel estimate.

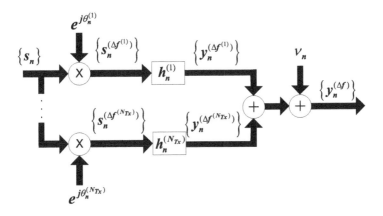

**Figure 6.1: Equivalent transmitter plus channel.**

Efficient channel estimation techniques are crucial to achieve reliable communication in wireless communication systems, and several techniques for ensuring accurate channel estimates have already been proposed ([SDM10], [DLF08], [XW05]). The efficiency of the conventional estimation techniques can eventually be enhanced with resort to the method proposed in [WY12], offering a good trade-off between the estimation performance and the computational complexity.

It is important to note that the SFN transmission creates severe artificial multipath propagation conditions. In order to mitigate its effects SFN systems employ a large number of OFDM subcarriers to ensure that the guard interval is large enough to cope with the maximum delay spread that can be handled by receivers. Albeit the system is defined to accommodate the worst-case scenario (which is given by the maximum delay spread), it also may represent a waste of bandwidth and excess of redundant information, in most cases. In [WWC⁺09], the channel length estimation problem is studied and the authors propose an autocorrelation-based algorithm to estimate the channel length without the need for pilots or training sequences. In order to improve spectral efficiency in the channel estimation, various methods that take advantage of the sparse nature of the equivalent CIR are presented in Chapter 5. In [WY12] [WPW08] are employed blind receivers, which although they do not need training sequences, may lead to performance degradation.

## 6.3.1   Frame Structure

In the following we will show that for a static scenario,[2] the knowledge of the CIR for each transmitter at the beginning of the frame, together with

---

[2]It should be emphasized that the equivalent CIR is not constant for static propagation conditions when we have different CFOs.

the knowledge of the corresponding CFO, is enough for obtaining the evolution of the equivalent CIR along the frame. The different CIRs and CFOs can be obtained by employing the frame structure of Fig. 6.2 (this structure allows us to track the evolution of the equivalent CIR along the frame, and it employs training sequences with the objective of knowing the CIR for each transmitter, as well as the corresponding CFO). We start by admitting that

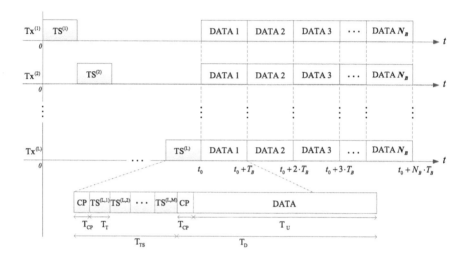

**Figure 6.2: Frame structure.**

the transmission of the training sequences is based on a scheduling scheme: each transmitter sends its training sequence $TS$, and then remains idle during the rest of the time slots reserved for training sequences transmission. Each training sequence includes a cyclic prefix whose duration $T_{CP}$ is longer than the duration of the overall channel impulse response (including the channel effects and the transmit and receive filters). The cyclic prefix is followed by $M$ (sub)blocks of size $N_T$ and duration $T_T$, which are appropriate for channel estimation purposes (e.g., based on Chu sequences or similar [DKFB04,DLF07]). The overall training sequence duration is $T_{TS} = T_{CP} + MT_T$.

Now, consider the $m^{th}$ (sub)block of the training sequence corresponding to the $l^{th}$ transmitter, $TS^{(l,m)}$. Using the corresponding samples the CIR can be obtained, and eventually enhanced using the sparse channel estimation techniques of Chapter 5, leading to the CIR estimates $\tilde{h}_n^{(l,m)}$, given by

$$\tilde{h}_n^{(l,m)} \approx h_n \cdot e^{j2\pi\Delta f^{(l)}mT_T} + \epsilon_n^{(l)}, \tag{6.8}$$

where the channel estimation error $\epsilon_n^{(l)}$ is Gaussian-distributed, with zero-mean. Note that the CIR estimates given by (6.8) are represent as

$$\tilde{h}_n^{(l,m)} \approx \tilde{h}_n^{(l,m-1)} \cdot e^{j2\pi\Delta f^{(l)}T_T}.\tag{6.9}$$

This means that we can obtain an estimate of $\Delta f^{(l)}$ from

$$\hat{\Delta} f^{(l)} \approx \frac{1}{2\pi T_T}\arg\left(\sum_{m=2}^{M}\sum_{n=1}^{N_T}\tilde{h}_n^{(l,m)}\tilde{h}_l^{(l,m-1)*}\right).\tag{6.10}$$

By compensating the phase rotation on each CIR estimate, an enhanced CIR estimate for the $l^{th}$ transmitter can be obtained, as follows:

$$\hat{h}_n^{(l)} = \frac{1}{M}\sum_{m=1}^{M}\tilde{h}_n^{(l,m)}\cdot e^{-j2\pi\hat{\Delta}f^{(l)}mT_T}.\tag{6.11}$$

One may think that a weakness of the proposed frame structure lies in a very long size when there are many transmitters, which causes inefficiency since a large portion of the training sequences remains idle. However, it is important to point out that although the length of the training increases with the number of transmitters (and a portion of the training remains idle for each transmitter), the inefficiency is not significant for the following reasons:

1. The number of relevant transmitters covering a given area is in general small (typically $L = 2$ or $L = 3$).

2. The frame associated with a given training interval can be very long, provided that there are accurate CFO estimates and the oscillators are reasonably stable. It is possible to have frames with several tens of data blocks.

3. The training block associated with each transmitter can have a duration much lower than data blocks.

Therefore, the efficiency can be very high.

## 6.3.2 Tracking the Variations of the Equivalent Channel

Assume that the channel remains unaltered within a block, only varying along the frame and that the frequency error is constant during the frame transmission interval. In these conditions, the information about the CFO of each transmitter allows us to track the variations of the equivalent channel. This means that it is possible to estimate the channel's impulse response for any time slot: the channel's impulse response at the instant $t_p$, (given by $\hat{h}_n^{(l)}(t_p)$), is the channel's impulse response at the initial instant 0 (given by $\hat{h}_n^{(l)}(0)$), multiplied by the phase rotation along that time interval (see Fig. 6.3).

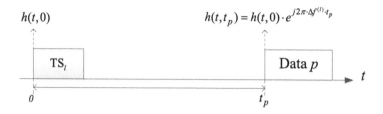

**Figure 6.3: Channel estimation for the $p^{th}$ block of data.**

This way, the CIR of the $l^{th}$ transmitter at the specific instant $t_p$ will be given by

$$\hat{h}_n^{(l)}(t_p) = \hat{h}_n^{(l)}(0)e^{j2\pi \cdot \hat{\Delta} f^{(l)} \cdot t_p},$$ (6.12)

with the equivalent CIR for the $p^{th}$ block of data given by

$$\hat{h}_n(t_p) = \sum_{l=1}^{N_{Tx}} \hat{h}_n^{(l)}(t_p) = \sum_{l=1}^{N_{Tx}} \hat{h}_n^{(l)}(0)e^{j2\pi \hat{\Delta} f^{(l)} t_p}.$$ (6.13)

Hence, the corresponding channel frequency response $\tilde{H}_k(t_p)$ can easily be obtained from $\tilde{h}_n(t_p)$.

## 6.4 Adaptive Receivers for SFN with Different CFOs

In the following, three frequency domain receivers are proposed for a non-synchronized SFN broadcasting system . For the sake of simplicity, an SFN transmission with two asynchronous transmitters will be considered, in which each transmitter is affected by a different CFO and the number of relevant transmitters covering a given area is generally small, typically $L = 2$ or $L = 3$. This, however, can be easily extended to a larger network, with more unsynchronized transmitters.

### 6.4.1 Method I

This receiver is entirely based on the IB-DFE. However it uses the initial CIR and CFO estimates provided by training sequences to estimate the equivalent channel, and updates the phase rotation for each data block of the frame. Nevertheless, this method does not perform CFO compensation, and it also assumes a constant equivalent channel within each block.

## 6.4.2  Method II

The corresponding receiver is illustrated in Fig. 6.4 and requires a small modification to the IB-DFE. It is developed from Method I, where after the phase update is performed the compensation of the average phase rotation, associated with the average CFO over the different transmitters. It considers the equivalent channel (given by (6.2)), in which the received signals associated with the $N_{Tx}$ transmitters are added, leading to the signal $\{y_n^{(\Delta f)}\}$.

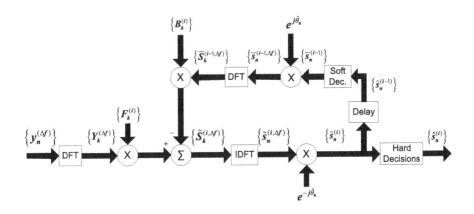

**Figure 6.4: Receiver structure for Method II.**

Instead of using the average phase rotation, a simple method based on the phase rotation associated with the strongest channel could be employed. However, since a different phase rotation is associated with each channel, an average phase compensation is more appropriate. The CFO compensation technique is based on a weighted average, in order to combine average values from samples corresponding to the CFOs associated with the different transmitters. The power of the channel associated with the $l^{th}$ transmitter is

$$P_{Tx}^{(l)} = \sum_{n=0}^{N-1} |h_n^{(l)}|^2 = \frac{1}{N} \sum_{k=0}^{N-1} |H_k^{(l)}|^2, \tag{6.14}$$

which means that the strongest channel has an higher contribution on the equivalent CFO. As a result, the equivalent CFO value, is given by

$$\hat{\Delta} f = \frac{\sum_{l=1}^{N_{Tx}} P_{Tx}^{(l)} \Delta f^{(l)}}{\sum_{l'=1}^{N_{Tx}} P_{Tx}^{(l')}}, \tag{6.15}$$

and therefore the average phase rotation is written as

$$\hat{\theta}_n = 2\pi\hat{\Delta}f\frac{n}{N}.$$

After compensating the average phase rotation of the received signal, the resulting samples are passed through a feedback loop in order to perform the equalization process.

### 6.4.3 Method III

It is important to note that Method II works well when the dispersion on the CFOs is not high. However, it is not efficient in the presence of substantially different CFOs. For instance, for two equal power transmitters with symmetric CFOs then equivalent CFO results in $\hat{\Delta}f = 0$ and no compensation is performed.

In Method III, a receiver that tries to jointly compensate the frequency offset associated with each transmitter and equalize the received signal is proposed. The objective is to use the data estimates from the previous iteration to obtain an estimate of the signal components associated with each transmitter, and posteriorly compensate the corresponding CFO. It is worth mentioning that for the first iteration the process is very straightforward, since there are no data estimates, and therefore, for the first iteration this receiver is reduced to a simpler version close to the one of Method II. For this reason, the feedback operations shown in Fig. 6.5, only apply to the subsequent iterations. The set of operations described next are performed for all $N_{Tx}$ signals within each iteration. Let us look to the $i^{th}$ iteration: the first operation consists of a filtering procedure, which isolates the signal $y_n^{(\Delta f^{(l)})}$, corresponding to the $l^{th}$ transmitter, by removing the contributions of the interfering signals from the overall received signal $y_n^{(\Delta f)}$, as given by equation (6.16).

$$\begin{aligned}
y_n^{(\Delta f^{(l)})} &= y_n^{(\Delta f)} - \sum_{l' \neq l}^{N_{Tx}} y_n^{(\Delta f^{(l')})} = y_n^{(\Delta f)} - \sum_{l' \neq l}^{N_{Tx}} s_n^{(\Delta f^{(l')})} * h_n^{(l')} \\
&\approx y_n^{(\Delta f)} - \sum_{l' \neq l}^{N_{Tx}} \hat{s}_n e^{j\hat{\theta}_n^{(l')}} * h_n^{(l')} \approx y_n^{(\Delta f)} - \sum_{l' \neq l}^{N_{Tx}} \hat{y}_n^{(\hat{\Delta}f^{(l')})}.
\end{aligned} \tag{6.16}$$

The computation of these undesired signal components is based on the equalized samples at the FDE's output from the previous iteration, $\{\hat{S}_k^{(i-1)}; k = 0, 1, \ldots, N-1\}$.

The samples corresponding to the signal $\{y_n^{(\Delta f^{(l)})}; n = 0, \ldots, N-1\}$ are then passed to the frequency-domain by an $N$-point DFT, leading to the corresponding frequency-domain samples which are then equalized by an appropriate frequency-domain feedforward filter. The equalized samples are converted back to the time-domain by an IDFT operation leading to the block

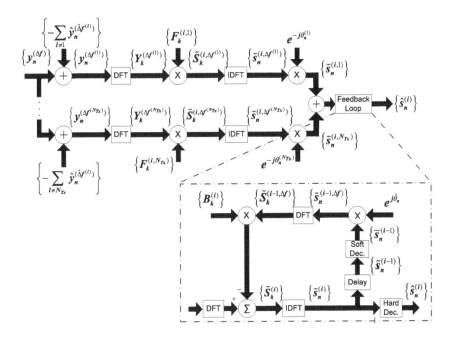

**Figure 6.5: Receiver structure for Method III.**

of time-domain equalized samples $\tilde{s}_n^{(i,\Delta f^{(l)})}$. Next, the resulting signal is compensated by the respective phase rotation $\theta_n^{(l)}$, which can easily be estimated from the original CIR and CFO estimates, as described before. This process is performed for each one of the $N_{Tx}$ signals, and the resulting signals are added in a single signal which is then equalized with resort to the feedback loop. The equalized samples at the FDE's output, are given by $\{\hat{S}_k^{(i)}; k = 0, 1, \ldots, N-1\}$, and for each iteration, the receiver compensates the phase error and combines the resulting signals before the feedback loop. The performance results in the next section will demonstrate that despite being more complex this receiver presents higher gains when compared to the first ones.

In terms of complexity, Method I and Method II have almost the same complexity as conventional receivers. However, Method III is slightly more complex since in each iteration it requires an additional FFT/IFFT pair for each branch (i.e., the number of FFT/IFFT pairs is proportional to $L$).

## 6.5 Performance Results

A set of performance results concerning the proposed frequency offset compensation methods for single frequency broadcast systems are presented next. It is assumed that identical signals emitted from different transmitters will arrive at the receiver with different delays, and will have different CIRs. Moreover, different CFOs between the local oscillator at each transmitter and the local oscillator at the receiver are considered. At the receiver's antenna, the signals are added being the result to consider over a single strong time-dispersive channel.

The chosen modulation relies on an SC-FDE scheme with blocks of $N = 4096$ subcarriers and a cyclic prefix of 512 symbols acquired from each block, although similar results were observed for other values of $N$, provided that $N \gg 1$. The modulation symbols belong to a QPSK constellation and are selected from the transmitted data according to a Gray mapping rule. For the sake of simplicity, linear power amplification at the transmitter was assumed. The performance results are expressed as a function of $E_b/N_0$, where $N_0$ is the one-sided power spectral density of the noise and $E_b$ is the energy of the transmitted bits.

Without loss of generality, it is considered an SFN transmission with two transmitters with different CFOs, where $\Delta f_{(1)}$ and $\Delta f_{(2)}$ denote the CFOs associated with the first and second transmitter, respectively. Another important parameter to be considered is the number $M$ of (sub)blocks following the cyclic prefix. In general the performance is different for different blocks, since the residual phase rotation on the signal associated with each transmitter increases as we move away from the training sequence or pilots (as the number of (sub)blocks increases). The subblock with worst performance is the one that is farthest from the training. In our simulations we considered frames with M=10 subblocks and the performance results concern the last subblock. However, it should be pointed out that for Method III with almost perfect CFO estimation the performance is almost independent of $M$.

Figs. 6.6 and 6.7 present the BER performance results for different values of $\Delta f^{(1)} - \Delta f^{(2)}$, namely from 0.05 to 0.175, for BER=$10^{-3}$. These results consider a difference of 10 dBs between the powers of the received signals from both transmitters $\left(\text{with } P_{Tx}^{(1)} > P_{Tx}^{(2)}\right)$. For comparison purposes, the results regarding the scenario in which the transmitters are not affected by CFO (i.e., $\Delta f^{(1)} = 0$ and $\Delta f^{(2)} = 0$) were also included. From the above performance results, it is clear that the transmission with non-synchronized transmitters can lead to significant performance degradation, particularly for Method I, where a very high deterioration of the BER performance with increasing values of $\Delta f^{(1)} - \Delta f^{(2)}$ can be observed. The reason for this is that this method does not perform a CFO compensation; it only updates the phase rotation for the channel associated with each transmitter for each block.

The curves obtained with resort to Method II show very reasonable results, since together with the IB-DFE iterations, this method performs the compensation of the average phase rotation (associated with the average CFO over the different transmitters). However, for high values of $\Delta f^{(1)} - \Delta f^{(2)}$ (typically $\approx 0.15$), it also indicates a significant degradation.

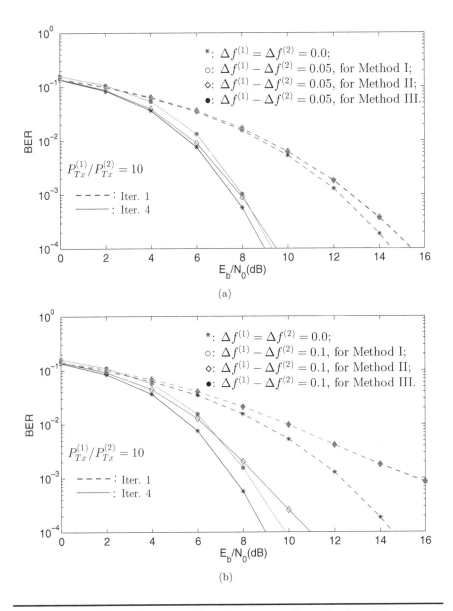

**Figure 6.6: BER performance for the proposed methods, with a power relation of 10dBs between both transmitters, and considering values of:** $\Delta f^{(1)} - \Delta f^{(2)} = 0.05$ **(a);** $\Delta f^{(1)} - \Delta f^{(2)} = 0.1$ **(b).**

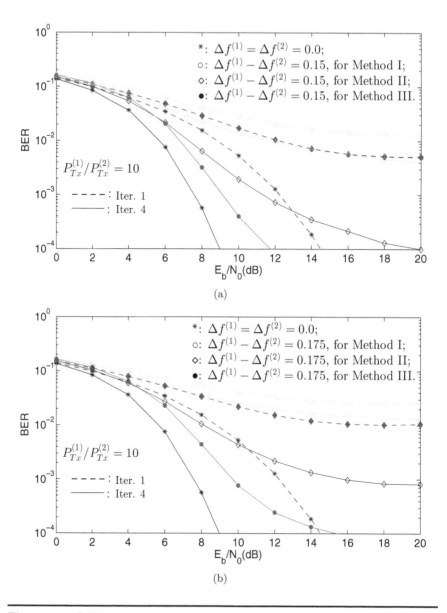

**Figure 6.7: BER performance for the proposed methods, with a power relation of 10dBs between both transmitters, and considering values of:** $\Delta f^{(1)} - \Delta f^{(2)} = 0.15$ **(a);** $\Delta f^{(1)} - \Delta f^{(2)} = 0.175$ **(b).**

In the performance results of Method III, it is clear that this method is capable of achieving very high gains, even with non-synchronized transmitters with strong values of $\Delta f^{(1)} - \Delta f^{(2)}$ (about 0.1). This method jointly compensates the frequency offset associated with each transmitter and equalizes the received signal; since it uses the data estimates from the previous iteration to obtain an estimate of the signal components associated with each transmitter, and posteriorly compensates the corresponding CFO. Despite being more complex, from the comparison of BER results for the fourth iteration it can be seen that for higher values of $\Delta f^{(1)} - \Delta f^{(2)}$ (about 0.15), Method III clearly surpasses Method II achieving a gain up to 8 dBs for $\Delta f^{(1)} - \Delta f^{(2)} = 0.15$ and about 5 dBs for $\Delta f^{(1)} - \Delta f^{(2)} = 0.175$. In order to provide a better analysis over the impact of the CFO on the performance, Figs. 6.8, 6.9, and 6.10 show the performance of the distinct methods regarding the different values of $\Delta f^{(1)} - \Delta f^{(2)}$ considered in the simulations. In the previous figures, we

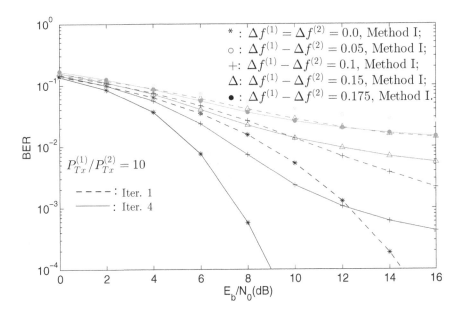

**Figure 6.8: Method I.**

presented the BER performance results for a relation of 10 dBs between the powers of the received signals from both transmitters. An interesting point of research would also be the influence of the received power on the performance associated with each one of the transmitters. In order to address this question, Figs. 6.11 and 6.12 present the performance results obtained with Method II and Method III, respectively, for different relations of the received power and $\Delta f^{(1)} - \Delta f^{(2)} = 0.15$. As shown in this figure, for both

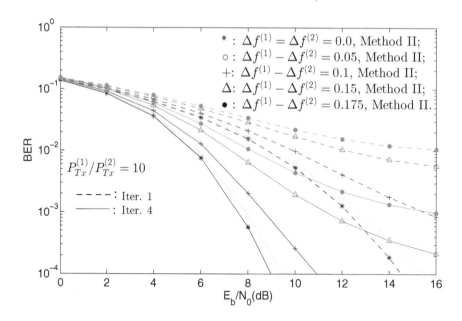

**Figure 6.9: Method II.**

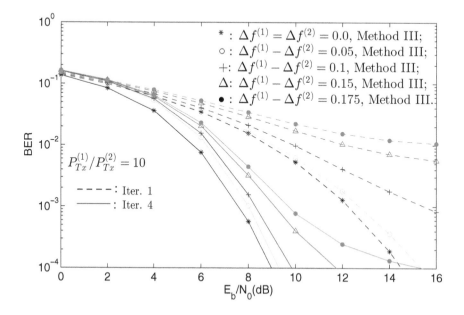

**Figure 6.10: Method III.**

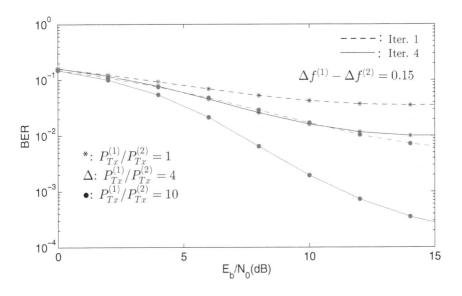

**Figure 6.11:** Impact of the received power on the BER performance, with $\Delta f^{(1)} - \Delta f^{(2)} = 0.15$, and employing the frequency offset compensation for Method II.

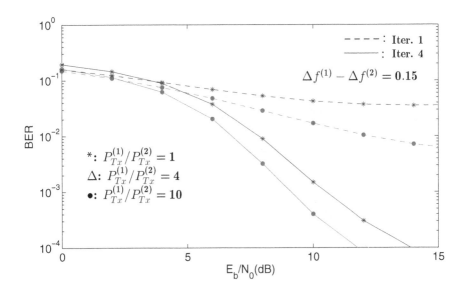

**Figure 6.12:** Impact of the received power on the BER performance, with $\Delta f^{(1)} - \Delta f^{(2)} = 0.15$, and employing the frequency offset compensation for Method III.

methods, the greater the difference in power between the transmitters (with $P_{Tx}^{(1)} > P_{Tx}^{(2)}$), the better the performance. This shows that the difference in power between transmitters has a strong impact on the system's performance. The above results show that the receivers based on the proposed methods are suitable for an SC-FDE scheme based on broadcasting transmission through an SFN system, even when transmitters have substantially different CFO.

Based on frequency offset compensation methods, these receivers consist of modified IB-DFE schemes, that equalize the received SC signal and compensate the residual CFOs. In order to achieve this, a frame structure was proposed with which it is possible to determine the channel's impulse response, as well as the CFO associated with each transmitter. That information makes a significant contribution to tracking variations of the equivalent channel during the frame duration. The performance results show significant gains on power efficiencies, especially when the receiver based on Method III is adopted. Therefore, despite the slight increase on the complexity of both receivers, they ensure excellent performance allowing good BERs in severely time-dispersive channels, even without perfect carrier synchronization between different transmitters.

# Chapter 7

# Multipath Channels with Strong Doppler Effects

In broadband mobile wireless systems the channel's impulse response can be very long leading to very large blocks, with hundreds or even thousands of symbols. Under these conditions it can be difficult to ensure a stationary channel during the block duration, which is a crucial requirement of conventional SC-FDE receivers. In order to avoid significant performance degradation due to strong Doppler effects, wireless systems based on SC-FDE schemes employ frequency-domain receivers which require an invariant channel within the block duration. Hence, a significant performance degradation occurs if the channel changes within the block's duration. SC-FDE detection is usually based on coherent receivers, therefore accurate channel estimates are mandatory. These channel estimates can be obtained based on training sequences and/or pilots [DGE01]. Although the use of training sequences allows an efficient and accurate channel estimation, as seen in Chapter 5, these estimates are local and the channel should remain almost constant between training blocks, something that might not be realistic in fast-varying scenarios due to strong Doppler effects.

The channel variations have different origins and effects. For instance, the previous chapter focused on the channel variations due to phase noise or residual CFO frequency errors, which can be a consequence of a frequency mismatch between the local oscillator at the transmitter and the local oscillator at the receiver. Nevertheless, this kind of channel variation leads to simple phase variations that are relatively easy to compensate at the receiver [SF08, DAPN10]. Another source of variation channel is the Doppler frequency shift caused by the relative motion between the transmitter and receiver. The channel variations due to this effect are not easy to compensate, and can become even more complex when the Doppler effects are distinct for different multipath components (e.g., when there exist different departure/arrival directions relative to the terminal movement). Therefore, it becomes mandatory to implement a tracking procedure to cope with channel variations between the training blocks. This can be done by employing decision-directed channel tracking schemes [MM96]. Detection errors might lead to serious error propagation effects. As an alternative, pilots multiplexed with data could be used for channel tracking purposes, as employed in most OFDM-based systems [HKR97]. Although adding pilots to OFDM systems is very simple (it just needs to assign a few subcarriers for that purpose), the same is not true for SC-FDE signals, where pilots lead to performance degradation and/or increased envelope fluctuations [LFDLD06, LFDL08]. Therefore, an efficient estimation and tracking schemes based on training blocks SC-FDE system is needed.

In this chapter various iterative receivers, able to attenuate the impact of strong Doppler effects, are proposed for SC-FDE schemes. Firstly, the short term channel variations are modeled as almost pure Doppler shifts which are different for each multipath component and use this model to design the frequency-domain receivers able to deal with strong Doppler effects. These receivers can be considered as modified turbo equalizers implemented in the frequency-domain, which are able to compensate the Doppler effects associated with different groups of multipath components while performing the equalization operation. The performance results will show that the proposed receivers have excellent performance, even in the presence of significant Doppler spread between the different groups of multipath components; this makes them suitable for SC-FDE scheme-based broadband transmission in the presence of fast-varying channels.

# 7.1   Doppler Frequency Shift due to Movement

Consider a transmission through a channel with multipath propagation, between a mobile transmitter traveling with speed $v$, and a fixed receiver, as shown in Fig. 7.1.

The relative motion between transmitter and receiver, induces a Doppler frequency shift in the received signal frequency, proportional to the speed of the transmitter, which depends on the spatial angle between the direction of the movement and the direction of departure/arrival of the component. Therefore, the Doppler shift associated with the $l^{th}$ multipath component is given by

$$f_D^{(l)} = \frac{v}{c} f_c \cos(\theta_l) = f_D^{max} \cos(\theta_l), \tag{7.1}$$

where $f_D^{max} = v f_c / c$ represents the maximum Doppler shift, proportional to

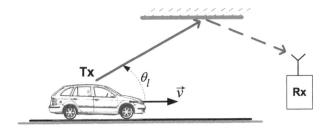

**Figure 7.1: Doppler shift.**

the vehicle speed $v$, $c$ denotes the speed of light, $\theta_l$ is the angle between $v$ and the arrival directions of the $l^{th}$ component.

## 7.2   Modeling Short-Term Channel Variations

The short-term channel variations are due to the receiver's motion [Rap01]. As the mobile moves over a short distance within a radio channel characterized by multipath fading, signal's power will vary rapidly originating small-scale fading due to the sum of many different multipath components, displaced with respect to one another in time and spatial orientation, having random amplitudes and phases. The received electromagnetic field at any point can be assumed to be composed of several horizontally traveling plane waves, having random amplitudes and angles of arrival for different locations. The amplitudes of the waves are assumed to be statistically independent, as well as the phases which are also uniformly distributed in $[0, 2\pi]$ [JC94]. Due to the fact that the different components have random phases the sum of the contributions exhibits a wide variation (e.g., even for small movements like a portion of a wavelength, the signal amplitude may vary by more than 40 dB).

Now, let $h(t, t_0)$ be the channel's impulse response associated with an impulse at time $t_0$ given by

$$h(t, t_0) = \sum_{l \in \Phi} \alpha_l(t_0)\delta\left(t - \tau_l\right), \tag{7.2}$$

where $\Phi$ is the set of multipath components, $\alpha_l(t_0)$ is the complex amplitude of the $l^{th}$ multipath component and $\tau_l$ its delay (without loss of generality, it is assumed that $\tau_l$ is constant for the short-term variations that are being considered). If the channel variations are due to Doppler effects we may write

$$\alpha_l(t_0) = \alpha_l(0)e^{j2\pi f_D^{(l)} t_0}. \tag{7.3}$$

Assuming the specific case were the receiver and all reflecting surfaces are fixed, and the transmitter is moving as shown in Figure 7.2(a)), therefore (7.2) can be rewritten as

$$h(t, t_0) = \sum_{l \in \Phi} \alpha_l(0) e^{j2\pi f_D^{(l)} t_0} \delta(t - \tau_l). \tag{7.4}$$

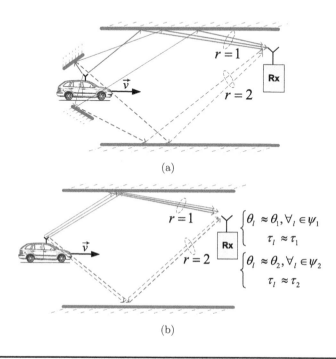

(a)

(b)

Figure 7.2: **Various objects in the environment scatter the radio signal before it arrives at the receiver (a); model where the elementary components at a given ray have almost the same direction of arrival (b).**

## 7.2.1 Generic Model for Short-Term Channel Variations

The following presents a generic model which considers a very high number of multipath components, especially when the reflective surfaces have a high roughness and / or have scattering effects. In order to overcome this problem, the model suggests that multipath components having the same direction of arrival (i.e., following a similar path), are grouped into clusters as shown in Figure 7.2(b). Under this approach, the overall channel will consist of the sum of individual time shifted channels, i.e.,

$$h(t, t_0) \simeq \sum_{r=1}^{N_R} \alpha_r(t_0) \delta(t - \tau_r), \qquad (7.5)$$

where $\alpha_r(t_0) = \sum_{l \in \Phi_r} \alpha_l(t_0)$, with $\Phi_r = \{l : \theta_l \simeq \theta^{(r)}\}$ denoting the set of element contributions grouped in the $r^{th}$ multipath group. Naturally, it means that $\tau_l \approx \tau_r$, $\forall_{l \in \Phi_r}$, i.e., the contributions associated with the $r^{th}$ multipath group have the same delay (at least at the symbol scale).

Due to the fact that $\alpha_l(t)$ is a random process depending on the path-loss and shadowing, whereas the phase factor $\phi_l(t)$ is a random process depending on the delay, among the Doppler shift and the carrier phase offset, $\alpha_l(t)$ and $\phi_l(t)$ can be considered as independent. Assuming the existence of a large number of scatterers within the channel, the CLT can be used to model the channel impulse response as a complex-valued Gaussian random process, and therefore allowing us to model the time-variant channel impulse response as a complex-valued Gaussian random process in the $t$ variable.

Hence, based on the CLT, $h_b(t, \tau)$ is approximately a complex Gaussian random process, and $\alpha_r(t_0)$ can then be regarded as a zero-mean complex Gaussian process with PSD characterized by

$$G_{\alpha_r}(f) \propto \begin{cases} \frac{1}{\sqrt{1-(f/f_D)^2}}, & |f| < f_D \\ 0, & |f| > f_D, \end{cases} \qquad (7.6)$$

which is depicted in Fig. 7.3(a) and corresponds to the so-called Jakes' Doppler spectrum. Thus, $\alpha_r(t_0)$ can be modeled as a white Gaussian noise $w(t_0)$, filtered by a filter with frequency response $H_D(f) \propto \sqrt{G_{\alpha_r}(f)}$, usually denoted "Doppler filter" [DB93].

## 7.2.2 A Novel Model for Short-Term Channel Variations

The generic model may not be suitable for broadband systems. The reason for that is simple: for narrowband systems the channel is modeled based on the assumption that the differences between the propagation delays among the several scattered signal components reaching the receiver are negligible when compared to the symbol period (i.e., the symbol duration is very high). The model then assumes that each multipath component following a given "macro path" is decomposed in several components (scattered at the vicinity of the transmitter). This is a fair approximation for narrowband systems. However, for broadband wireless mobile systems, multipath components that depart/arrive with substantially different directions will have delays that are very different and therefore they should not be regarded as elementary components of the same ray. This means that all elementary components at a given

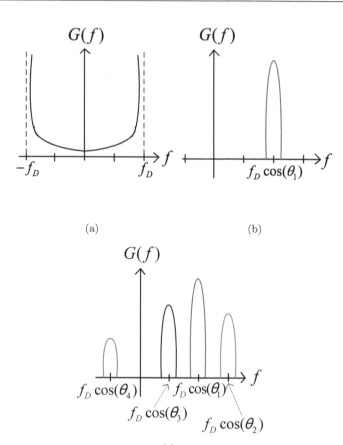

**Figure 7.3: Jakes PSD (a). PSD associated with the transmission of a single ray (b). PSD associated with the transmission of multiple rays (c).**

ray should have similar direction of departure/arrival. Therefore, the Doppler filter must have a very narrow bandwidth centered in $f_D^{(r)} = f_D \cos(\theta_r)$, and consequently, short-term channel variations can be modeled as almost pure Doppler shifts that are different for each multipath group, i.e.,

$$\alpha_r(t_0) \simeq \alpha_r(0)e^{j2\pi f_D \cos(\theta_r)t_0}, \tag{7.7}$$

(it is important to note that $\alpha_r(0)$ can still be modeled as a sample of a zero-mean complex Gaussian process). Under these conditions, the Doppler spectrum associated with each multipath group will have a narrow band nature, as depicted in Fig. 7.3(b). Fig. 7.3(c) illustrates the Doppler spectrum considering a set of different multipath groups.

The time-varying channel impulse response can then be written as

$$h(t, t_0) \simeq \sum_{r=1}^{N_R} h^{(r)}(t, 0)e^{j2\pi f_D^{(r)} t_0},\qquad(7.8)$$

where each individual channel $h^{(r)}(t, 0)$ is characterized by a normal PDP, representing the cluster of multipath components having a similar direction of arrival (although with substantially different delays), and is given by

$$h^{(r)}(t, 0) = \sum_{l \in \Phi_r} \alpha_l(0)\delta\left(t - \tau_l\right),\qquad(7.9)$$

where $\Phi_r$ denotes the set of all multipath components. Fig. 7.4 shows an example of the clustering process. Of course, in a practical scenario it might be necessary to perform a kind of quantization of the Doppler shifts.

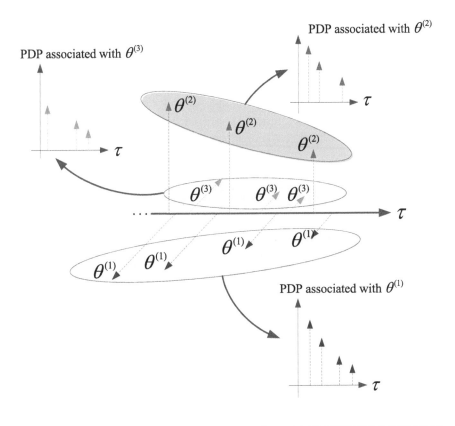

**Figure 7.4:** Multipath components having the same direction of arrival $\theta$ are grouped into clusters.

## 7.3 Channel Estimation and Tracking

As already pointed out, the present work assumes coherent receivers which require accurate channel estimates. The estimates can be obtained with the help of appropriate training sequences, or by employing the efficient channel estimation methods presented in Chapter 5, which take advantage of the sparse nature of the equivalent CIR.

In the following it will be shown that the knowledge of the CIR at the beginning of the frame, together with the knowledge of the corresponding Doppler drifts, is enough to obtain the evolution of the equivalent CIR along all the frame.

### 7.3.1 Channel Estimation

The different CIRs and Doppler drifts can be obtained by employing the frame structure of Fig. 7.5, that starts with the transmission of two training sequences, denoted $TS_1$ and $TS_2$, respectively. Each training sequence includes a cyclic prefix with duration $T_{CP}$, which is longer than the duration of the overall channel impulse response (including the channel effects and the transmit and receive filters), followed by the useful part of the block with duration $T_{TS}$, which is appropriate for channel estimation purposes.

Between the training sequences there is a period of time $\Delta T$, which may be available for data transmission. By employing high values of $\Delta T$ the accuracy of the estimates can be significantly improved, but it should be assured that the phase rotation within this time interval $\Delta T$ does not exceed $\pi$.

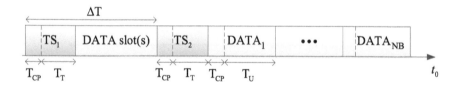

**Figure 7.5: Frame structure.**

Consider the first training sequence, $TS_1$. From the corresponding samples it is possible to obtain the CIR, which can be eventually enhanced with resort to the sparse channel estimation techniques of Chapter 5, leading to the set of CIR estimates $\tilde{h}_n^{(1)}$. It can be easily shown that the corresponding estimates can be given by

$$\tilde{h}_n^{(1)} = h_n(0) + \epsilon_n^{(1)}, \tag{7.10}$$

where the $h_n(0)$ represents the initial impulse response channel, and estimation error $\epsilon_{n,l}$ is Gaussian-distributed, with zero-mean. Now consider the second training sequence,

$$\tilde{h}_n^{(2)} = h_n(\Delta T) + \epsilon_n^{(2)}, \tag{7.11}$$

where $h_n(\Delta T)$ denotes the channel impulse response obtained at the instant $\Delta T$ and it is simply the initial impulse response $h_n(0)$ times the corresponding phase rotation. Naturally, this is only applicable to relevant multipath components (i.e., the multipath components power must exceed a pre-defined threshold, otherwise the samples are considered noise and ignored). Therefore, (7.11) can be rewritten as

$$\tilde{h}_n^{(2)} = h_n(0) \cdot e^{j2\pi f_D \cos(\theta_n)\Delta T} + \epsilon_n^{(1)}. \tag{7.12}$$

The channel evolution between these training blocks can be obtained from the parameters which characterize each multipath component, as will be explained next.

### 7.3.2 *Tracking of the Channel Variations*

We have seen that the short-term time variations of a mobile radio signal (which are a consequence of the transmitter (or receiver) motion in space [Rap01]) can be directly related to the corresponding time-varying channel impulse response. Let us then consider a specific case where the receiver and all reflecting surfaces are fixed, and the transmitter is moving. In these conditions, variations on the mobile channel are due to Doppler effects, and are given by

$$\alpha_l(t_0) = \alpha_l(0)e^{j2\pi f_l t_0} \tag{7.13}$$

and in these conditions (7.2) can be rewritten as

$$h(t, t_0) = \sum_{l \in \Phi} \alpha_l(0)e^{j2\pi f_l t_0}\delta(t - \tau_l), \tag{7.14}$$

It is therefore important to be able to predict the channel response for transmission within fast-varying scenarios.

#### 7.3.2.1 *Using the Sampling Theorem to Track the Channel Variations*

A precise tracking of the channel variations can be derived from a direct application of the sampling theorem: as was shown before, if it is admitted that the channel is characterized by a Doppler spectrum, then the channel can be seen as if $\alpha_l(t_0)$ had been modeled as a white Gaussian noise $w(t_0)$, filtered by a filter with frequency response $H_D(f) \propto \sqrt{G_{\alpha_l}(f)}$, with the Doppler spectrum occupying a bandwidth $f_D$ (corresponding to the maximum Doppler frequency). Sampling $\alpha_l(t_0)$ at a rate $R_a \geq 2f_D$, results in the set $\{\alpha_l(nT_a)\}$ which is statistically sufficient for obtaining $\alpha_l(t_0)$. Despite being a very straightforward process, this might lead to implementation difficulties due to the data storage and delays inherent in channel interpolation, especially when the training blocks are transmitted at a rate close to $2f_D$.

### 7.3.2.2  A Novel Tracking Technique

In this section is proposed an efficient channel tracking technique for SC-FDE transmission over fast-varying multipath channels. Instead of modeling the channel as a random process with bandwidth $f_D$, a different approach is followed by modeling the individual multipath components as time-varying signals characterized by fixed parameters (e.g., the Doppler drift of each individual multipath component). In order to do this, a method for estimating the parameters that characterize each multipath component is employed. These parameters are then used for obtaining the channel evolution between training blocks that are transmitted with a rate much lower than $2f_D$. In these conditions it can be considered that the channel evolution is not random but, in fact, completely deterministic.

First is presented the method for estimating the parameters that characterize each multipath component, which can then be used to obtain the channel evolution between training blocks that are transmitted with a rate $F_a \ll 2f_D$. Regarding the $l^{th}$ component these parameters are: the complex amplitude $\alpha_l(t)$, delay $\tau$, direction of arrival $\theta_l$, and the Doppler drift $f_l = f_D \cos(\theta_l)$.

The process is very simple: by knowing the initial value of the complex amplitude, $\alpha_l(0)$, which can be acquired from the estimation of $h_n(0)$, and assuming that all the other parameters are fixed (which is reasonable since we are assuming broadband systems) it can be admitted that the channel evolution is completely deterministic.

To better understand this, regard the frame structure proposed in Fig. 7.5. The parameter $\alpha_l(0)$ can be acquired from the estimation of $h_n(0)$ with resort to the training sequence $TS_1$. In the same way, the value of the complex amplitude, $\alpha_l(\Delta T)$, can be acquired with resort to $TS_2$. From (A.9) it is clear that the difference between $\alpha_l(0)$ and $\alpha_l(\Delta T)$ is due to the phase rotation related to Doppler effects, along the time interval $\Delta T$. Therefore, the equivalent Doppler shift corresponding to the $l^{th}$ multipath component, can be obtained from

$$
\hat{f}_l = \frac{1}{2\pi\Delta T} \arg\left(\alpha_l(\Delta T) \cdot \alpha_l(0)^*\right)
$$

$$
\approx f_l + \frac{\varepsilon_l^Q}{2\pi \cdot \Delta T \cdot |\alpha_l(0)|^2},
$$

(7.15)

where $\varepsilon_l^Q$ represents the quadrature component of the noise contribution.

Still, it is important to guarantee that $|\alpha_l(0)| \approx |\alpha_l(\Delta T)|$, otherwise it may become necessary to increase the power of the training blocks, or to employ more sophisticated estimation techniques. Naturally, this is only applicable to relevant multipath components (i.e., the multipath components whose power exceeds a pre-defined threshold. Any samples below this limit are considered noise and ignored).

Regard the estimator's variance, given by

$$\sigma^2_{\hat{f}_l} \simeq \frac{\sigma^2_{\varepsilon^Q_l}}{(2\pi \cdot \Delta T \cdot |\alpha_l(0)|^2)^2}.$$
(7.16)

If $|\alpha_l(0)|^2 \gg \sigma^2_{\varepsilon^Q_l}$, the noise contribution will be insignificant and we can have a high precision estimate of the Doppler drift. Hence, for the $l^{th}$ multipath component the knowledge of the initial value of the complex amplitude $\alpha_l(0)$, along with the corresponding Doppler shift $f_l$, allows us to track the variations of the channel's impulse response for any slot of the frame along that time interval.

## 7.4 Receiver Design

Let us now consider an SC-FDE transmission system through a multipath channel with strong Doppler effects. We assume that each cluster of rays is associated with a different frequency drift due to Doppler effects, and we present two methods to compensate these effects at the receiver side. Under these conditions, each sample is affected by a different frequency drift. For an SC-FDE system the frequency drift induces a rotation in the constellation that grows linearly along the block. Without loss of generality, we assume a null phase rotation at the first sample $n = 0$.

In [DAPN10], an estimation and compensation technique of the phase rotation associated with the frequency drift is proposed for a conventional cellular system in a slowly varying scenario. Nevertheless, the multipath propagation causes time dispersion, and multiple sets (clusters) of rays received with different delays are added in the receiver. Moreover, for fast-varying channels the received signal will fluctuate within each block. Therefore, regarding these conditions, it is admitted that the received signals arrive with possible different delays, and are exposed to different frequency drifts. It is also assumed that the multipath components with similar Doppler frequency shift $f_l$ are grouped into clusters, and a method to compensate these effects at the receiver side is presented. Therefore, in time domain the received equivalent block, $y_n^{(f_D)}$, will be the sum of the time-domain blocks associated with the $N_R$ sets of rays, as follows

$$y_n^{(f_D)} = \sum_{r=1}^{N_R} y_n^{(r)} e^{j2\pi f_D^{(r)} n/N},$$
(7.17)

where $f_D^{(r)}$ denotes the Doppler drift associated with the $r^{th}$ cluster of rays. Let

$$\theta_n^{(r)} = 2\pi f_D^{(r)} \frac{n}{N},$$
(7.18)

then (7.17) can be rewritten as

$$y_n^{(f_D)} = \sum_{r=1}^{N_R} y_n^{(r)} e^{j\theta_n^{(r)}}. \tag{7.19}$$

Under these conditions the transmitter chain associated with each one of the $N_R$ cluster of rays can be modeled as shown in Fig. 7.6. Considering a transmission associated with the $r^{th}$ cluster of rays, in the presence of a Doppler drift $f_D^{(r)}$, the block of time-domain data symbols is affected by a phase rotation (before the channel), resulting in the effectively transmitted block, $\{s_n^{(f_D^{(r)})}; n = 0, ..., N - 1\}$. It follows from (7.19) that the Doppler drift induces a rotation $\theta_n^{(r)}$ in block's symbols that grows linearly along the time-domain block. Obviously, the effect of this progressive phase rotation might lead to a significant performance degradation.

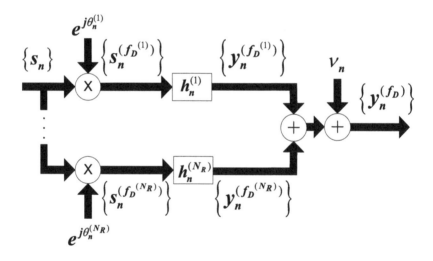

**Figure 7.6: Equivalent cluster of rays plus channel.**

In the following are proposed two frequency domain receivers, based on the IB-DFE, with joint equalization and Doppler drift compensation. The first receiver whose structure is depicted in Fig. 7.7 has small modifications compared to the IB-DFE, and employs joint equalization and Doppler drift compensation. It considers the equivalent channel, in which the received signals associated with the $N_R$ sets of rays are added leading to the signal $y_n^{(f_D)}$. To perform the Doppler drift compensation, one could employ a simple method based on the fact that the equivalent frequency drift, $\hat{f}_D$, corresponds to the one (pre-

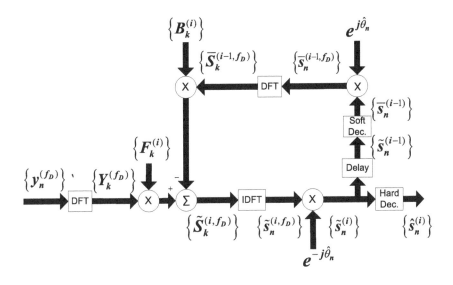

**Figure 7.7: Receiver structure for ADC.**

viously estimated) associated with the strongest subchannel. However, each cluster suffers a different phase rotation, so an average phase compensation is more appropriate. Thus, for this iteration, the Doppler drift compensation technique is based on a weighted arithmetic mean, in order to combine average values from samples corresponding to the frequency drifts associated with the different clusters. The average power associated with each cluster is denoted by

$$P^{(r)} = \sum_{n=0}^{N-1} |h_n^{(r)}|^2 = \frac{1}{N} \sum_{k=0}^{N-1} |H_k^{(r)}|^2, \qquad (7.20)$$

and it is easy to see that the strongest subchannel will have a higher contribution to the equivalent frequency drift. As result, the estimated frequency offset value and the estimated phase rotation are given by

$$\hat{f}_D = \frac{\sum\limits_{r=1}^{N_R} P^{(r)} f_D^{(r)}}{\sum\limits_{r'=1}^{N_R} P^{(r')}}, \qquad (7.21)$$

and

$$\hat{\theta}_l = 2\pi \hat{f}_D \frac{n}{N}, \qquad (7.22)$$

respectively. After the compensation of the estimated phase rotation affecting

the received signal, the resulting samples are passed through a feedback operation in order to complete the equalization procedures. The Doppler drift compensation technique employed in this receiver can be called average Doppler
compensation (ADC). However, the fact that it is based on an average phase
compensation might have implications in performance.

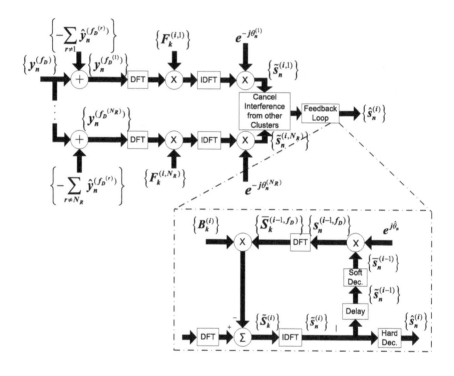

**Figure 7.8: Receiver structure for TDC.**

Consider now the second receiver shown in Fig. 7.8. This receiver employs
a Doppler drift compensation technique called total Doppler compensation
(TDC), which compensates the Doppler drift associated with each cluster of
rays individually. It is worth mentioning that for the first iteration the process
is equivalent to a linear receiver due to the absence of data estimates. Only for
the subsequent iterations, this receiver will jointly compensate the estimated
phase rotation due to Doppler drift and equalize the received signal. Hence,
the feedback operations which will be described next are only valid for the
subsequent iterations.

Regard the received signal referring to the $r^{th}$ cluster of rays, given by

$$
\begin{aligned}
y_n^{(f_D^{(r)})} &= y_n^{(f_D)} - \sum_{r' \neq r}^{N_R} y_n^{(f_D^{(r')})} = y_n^{(f_D)} - \sum_{r' \neq r}^{N_R} s_n^{(f_D^{(r')})} * h_n^{(r')} \\
&\approx y_n^{(f_D)} - \sum_{r' \neq r}^{N_R} \hat{s}_n e^{j\hat{\theta}_n^{(r')}} * h_n^{(r')} \approx y_n^{(f_D)} - \sum_{r' \neq r}^{N_R} \hat{y}_n^{(f_D^{(r')})},
\end{aligned}
\tag{7.23}
$$

where $*$ denotes the convolution operation, and $\hat{y}_n^{(f_D^{(r')})}$ represents the estimates of the received signal components. The set of operations described next are performed for all $N_R$ signals within each iteration: the first operation consists of isolating from the total received signal $y_n^{(f_D)}$ the signal associated with the $r^{th}$ cluster of rays $y_n^{(f_D^{(r)})}$, which is accomplished by removing the contributions of the interfering signals as described in (7.23). The computation of the undesired signal components is based on the data estimates at the FDE's output from the previous iteration, $\{\hat{S}_k^{(i-1)}; k = 0, 1, \ldots, N-1\}$. The samples corresponding to the resulting signal $\{y_n^{(f_D^{(r)})}; n = 0, \ldots, N-1\}$ are then passed to the frequency-domain by an $N$-point DFT, leading to the corresponding frequency-domain samples which are then equalized by a frequency-domain feedforward filter. The equalized samples are converted back to the time-domain by an IDFT operation leading to the block of time-domain equalized samples $\hat{s}_n^{(f_D^{(r)})}$. Next, the Doppler drift of the resulting signal is compensated by the respective estimated phase rotation $\theta_n^{(r)}$, which for simplicity is assumed to have been previously estimated. This process is performed for each one of the clusters of multipath components, and the signals are added in a single signal which is then equalized with resort to the IB-DFE. The equalized samples at the FDE's output will be given by $\{\hat{S}_k^{(i)}; k = 0, 1, \ldots, N-1\}$. Therefore, the receiver jointly compensates the phase error and equalizes the received signal by a Doppler drift compensation before the equalization and detection procedures.

## 7.5   Performance Results

Here is presented a set of performance results regarding the use of the proposed receiver in time-varying channels.

An SC-FDE modulation is considered, with blocks of $N = 1024$ symbols and a cyclic prefix of 256 symbols acquired from each block (although similar results were observed for other values of $N$, provided that $N \gg 1$). The modulation symbols belong to a QPSK constellation and are selected from the transmitted data according to a Gray mapping rule. Linear power amplification at the transmitter is also assumed.

For each multipath group, the Doppler drift and the respective channel impulse response are obtained with the help of the frame structure presented previously in Sec. 7.3.1.

Firstly, consider the scenario of Fig. 7.9 where the receiver and all reflecting surfaces are fixed, and the transmitter (i.e., mobile terminal) is moving with speed $v$. The channel is admitted to have uncorrelated Rayleigh fading, with multipath propagation, and with short-term variations due to Doppler effects. The maximum normalized Doppler drift is given by $f_d = f_D T_B = v \frac{f_c}{c} T_B$, with $f_c$ denoting the carrier frequency, $c$ the speed of light and $T_B$ the block duration.

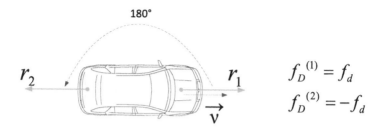

**Figure 7.9: Transmission scenario with two clusters of rays.**

Consider now a critical scenario, where the multipath components are divided into two multipath clusters. The first cluster has the direction of movement and therefore is associated with a Doppler drift of $f_D^{(1)} = f_d$, while the second group has the opposite direction and a Doppler drift of $f_D^{(2)} = -f_d$. Without loss of generality it is assumed that 64 multipath components arrive from each direction, and there is a difference of 10 dBs between the powers of both clusters $\left(\text{with } (P^{(1)} > P^{(2)})\right)$. Figs. 7.10 and 7.11 present the BER performance for the proposed methods where ADC and TDC are denoted as Method I and Method II, respectively, regarding a transmission with a maximum normalized Doppler drift of $f_d = 0.05$ and $f_d = 0.09$. For comparison purposes the results for a static channel are also included. Regarding the results, both compensation methods ADC and TDC, together with the IB-DFE iterations, can achieve high power gains, even with strong values of Doppler drifts. The two methods' performance is almost the same for BER values higher than $10^{-2}$. For both scenarios, their performance is very good when compared with the SC-FDE without compensation (see results for one iteration). As can be seen from Fig. 7.10 at BER of $10^{-3}$, the performance of both methods outperforms the SC-FDE by more than 6 dB. In fact the proposed compensation methods can achieve higher power efficiency, even in

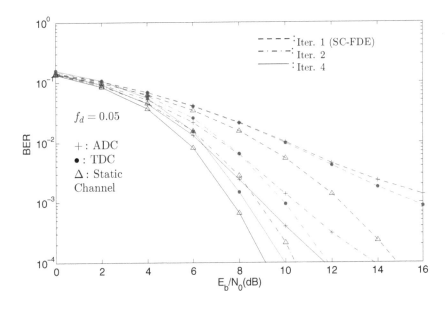

**Figure 7.10: BER performance for a scenario with normalized Doppler drifts $f_d$ and $-f_d$ for $f_d = 0.05$.**

the presence of several groups of rays with significant differences in Doppler drifts. The TDC method gives the best error performance at the expense of computational complexity. Despite being more complex, for moderate values of Doppler drifts ($f_d \approx 0.05$), it outperforms the ADC method by 1.75 dB for a BER of $10^{-4}$. For higher values of Doppler drifts, i.e., $f_d \approx 0.09$, the method TDC overcomes method ADC (whose BER performance highly deteriorates), achieving a gain of several dBs over the ADC method. For instance, from Fig. 7.11, for the $4^{th}$ iteration at BER of $10^{-3}$ the power gain is near to 7 dB. Again, we see that the TDC method performs very well and provides a good tradeoff between the error performance and the decoding complexity when compared with the ADC method. Moreover, for moderate Doppler drifts it can be seen from Fig. 7.10 that the second method's performance is close to the static channel (with a power degradation lower than 1 dB). Therefore, and despite the increase in complexity, the receiver based on the second method has excellent performance, even when the different clusters of multipath components have strong Doppler effects.

From our performance results we may conclude that the proposed compensation methods can achieve high gains, even for several groups of rays with substantially different Doppler drifts. Therefore, the proposed receivers are

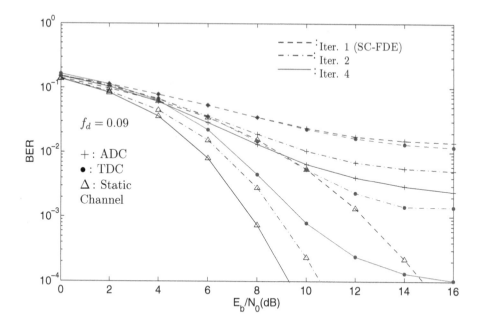

**Figure 7.11: BER performance for a scenario with normalized Doppler drifts $f_d$ and $-f_d$ for $f_d = 0.09$.**

suitable for SC-FDE transmission, and can have excellent performance in the presence of fast-varying channels.

# Appendix A

# Important Statistical Parameters

Given a continuous random variable $R$, let us consider the event $\{R \leq r\}$, with $r$ representing a real number in $[-\infty, \infty]$. The probability that this event will occur can be written as $P(R \leq r)$. This probability can be defined by a function $F_R(r)$, defined as

$$F_R(r) = P(R \leq r) = \int_0^R f_R(r) dr = 1 - e^{\frac{-r^2}{2\sigma^2}}, (-\infty < r < \infty). \qquad (A.1)$$

This function is called *cumulative distribution function* (CDF) or *probability distribution function* of the random variable $R$, and gives the probability of a random variable $R$ (for instance representing the envelope's signal) not exceeding a given value $r$. A continuous random variable has a continuous distribution function, and this function is right-continuous, increasing monotonically. And since $F_R(r)$ is a probability, it is limited to the values $F_R(-\infty) = 0$ and $F_R(\infty) = 1$, in the interval $0 \leq F_R(r) \leq 1$.

Since the CDF $F_R(r)$ is a continuously differentiable function of $r$, its derivative function is called the *probability density function* (PDF) of the random variable $R$. The PDF can be defined as

$$f_R(r) = \frac{dF_R(r)}{dr}, (-\infty < r < \infty). \qquad (A.2)$$

Due to the fact that $F_R(r)$ is a non-decreasing function of $r$, it follows that $f_R(r)$ is non-negative since $f_R(r) = \dfrac{dF_R(r)}{dr} = \lim\limits_{\Delta r \to \infty} \dfrac{F_R(r + \Delta r) - F_R(r)}{\Delta r} \geq 0$ is valid for all $r$. (In the same way, the integration of this density function

**131**

results in the corresponding cumulative distribution function). The CDF gives the area under PDF from $-\infty$ to $r$.

Other very important statistical parameters are the mean value, the mean-square value, the variance, and the standard deviation.

The mean value (or expected value or statistical average)[1] of a random variable $R$ is written as $\mathbf{E}[R]$, where $\mathbf{E}$ denotes the statistical expectation operator. Considering a continuous R.V., characterized by a probability density function, the mean of $R$ is given by

$$\mathbf{E}[R] = \int_{-\infty}^{\infty} r f_R(r) dr. \tag{A.3}$$

In turn, the mean-square value is given by

$$\mathbf{E}[R^2] = \int_{-\infty}^{\infty} r^2 f_R(r) dr. \tag{A.4}$$

The variance of a random variable $R$ is a measure of the concentration of $R$ around its expected value, and can be written as

$$\sigma_R^2 = \text{Var}(R) = \mathbf{E}[(R - \mathbf{E}[R])^2] = \int_{-\infty}^{\infty} (r - \mathbf{E}[R])^2 f_R(r) dr, \tag{A.5}$$

where $\text{Var}\{\cdot\}$ represents the variance operator.

The standard deviation, which represents the root mean-square value of the random variable $R$ around its expected value, is given by

$$\sigma_R = \sqrt{\text{Var}(R)} = \sqrt{\mathbf{E}[(R - \mathbf{E}[R])^2]}. \tag{A.6}$$

Another very important concept of statistics are the moments. In fact, the mean and variance can be written in terms of the first two moments $\mathbf{E}[R]$ and $\mathbf{E}[R^2]$. The $k^{th}$ moment of the random variable $R$ is given by

$$\mathbf{E}[R^k] = \int_{-\infty}^{\infty} r^k f_R(r) dr, k = 0, 1, ... \tag{A.7}$$

while the $k^{th}$ central moment is defined as

$$\mathbf{E}[(R - \mathbf{E}\{R\})^k] = \int_{-\infty}^{\infty} (R - \mathbf{E}\{R\})^k f_R(r) dr, k = 0, 1, ... \tag{A.8}$$

## A.1 Rayleigh Distribution

The Rayleigh distribution is widely employed in wireless channel modeling to describe the distribution of the received signal envelope when the LOS component does not exist.

---

[1]Depending on the type of variable, the mean value and the expected value may be the same.

Let us consider any two statistically independent Gaussian random variables $X_r$ and $X_n$, with zero mean and variance $\sigma^2$. A new random variable $R$, can be derived from $X_r$ and $X_n$ by doing

$$R = \sqrt{X_r^2 + X_n^2}, \tag{A.9}$$

where $R$ represents a Rayleigh distributed R.V., characterized by a Rayleigh distribution given by

$$F_R(r) = P(R \leq r) = \int_0^R f_R(r)dr = 1 - e^{\frac{-r^2}{2\sigma^2}}, r \geq 0. \tag{A.10}$$

The derivation of the CDF given by $F_R(r)$ yields the corresponding probability density function (PDF)

$$f_R(r) = \begin{cases} \frac{r}{\sigma^2} e^{\frac{-r^2}{2\sigma^2}}, & 0 \leq r < \infty, \\ 0, & r < 0, \end{cases} \tag{A.11}$$

which is known as Rayleigh PDF. Its mean value is given by

$$\mathbf{E}[R] = \int_0^\infty r f_R(r)dr = \sigma\sqrt{\frac{\pi}{2}} \tag{A.12}$$

The mean-square value, given by the second moment, is

$$\mathbf{E}[R^2] = \int_0^\infty r^2 f_R(r)dr = 2\sigma^2 = R_{rms}^2, \tag{A.13}$$

while the variance is given by

$$\mathrm{Var}(R) = \mathbf{E}[(R - \mathbf{E}[R])^2] = \mathbf{E}[R^2] - \mathbf{E}^2[R]$$
$$= 2\sigma^2 - (\sigma\sqrt{\frac{\pi}{2}})^2 \tag{A.14}$$
$$= \sigma^2 \left(2 - \frac{\pi}{2}\right)$$

## A.2  Rician Distribution

Rayleigh fading assumes that all incoming multipath components travel by relatively equal paths. However, and as often occurs in practice, in addition to the $N$ multipath components, the propagation channel is characterized by a strong, dominant stationary signal component (i.e., line-of-sight propagation path) [Rap01]. In this case the received signal is constituted by the superposition of a complex Gaussian component and a LOS component. In these cases, the Rician distribution is employed to model the statistics of the fading envelope. The Rician fading model and its analysis are equivalent to that of the Rayleigh fading case, but with the addition of a constant term. Hence, the signal envelope has a PDF described by the Rician distribution [Ric44] given by

$$f_R(r) = \begin{cases} \frac{r}{\sigma^2} e^{-\frac{(r^2+\nu^2)}{2\sigma^2}} I_0\left(\frac{r\nu}{\sigma^2}\right), & r \geq 0, \\ 0 & r < 0, \end{cases} \tag{A.15}$$

where the parameter $\nu$ represents the envelope of the stationary signal component (i.e., peak amplitude of the dominant signal) of the received signal, while $I_0$ denotes the zeroth-order modified Bessel function of the first kind, and $2\sigma^2$ denotes the power of the Rayleigh component.

A key factor in the model's analysis is given by the "Rician $K$-Factor" which is defined as the ratio between the deterministic signal power and the power of the multipath components.

$$K = \frac{\nu^2}{2\sigma^2}. \tag{A.16}$$

This factor is often expressed in dB by

$$K(dB) = 10 \log \frac{\nu^2}{2\sigma^2} \text{ dB.} \tag{A.17}$$

The parameter $K$ is a fundamental factor since it is able to completely specify the Ricean distribution, and it gives the ratio of the power in the LOS component to the power in the other multipath components. As the stationary signal component reduces its power, i.e., as $K \to 0$, $I_0(0) = 1$, the Rician PDF becomes a Rayleigh PDF. The reason for this is that as the stationary (dominant) signal component becomes weaker, the composite signal appears like a typical noise signal which in its turn is characterized by a Rayleigh envelope. if the stationary signal component is much higher than the random multipath components power (i.e., as $K \to \infty$), it can be assumed that only the LOS component is present, corresponding to a situation in which the channel is not affected by multipath fading. In this scenario, the Gaussian PDF represents a good approximation for the Rician pdf (i.e., the Ricean PDF is approximately Gaussian about the mean) [Rap01]. On the other hand, if the dominant signal fades away then the Ricean distribution turns into a Rayleigh distribution. Hence, the parameter $K$ can be seen as a fading measure since a small $K$ corresponds to severe fading, while a large $K$ leads to low fading.

The cumulative distribution function $F_R(r)$ can be given by

$$F_R(r) = \begin{cases} 1 - Q\left(\frac{\nu}{\sigma}\frac{r}{\sigma}\right), & r \geq 0 \\ 0 & r < 0 \end{cases} \tag{A.18}$$

with $Q$ denoting the *Marcum Q-function* given in [PM06]. The first two moments of the Rician distributed random variables $r$ can be given by

$$\mathbf{E}[R] = \sigma\sqrt{\frac{\pi}{2}} \, {}_1F_1\left(-\frac{1}{2}; 1; -\frac{\nu^2}{2\sigma^2}\right), \tag{A.19}$$

and

$$\mathbf{E}[R^2] = 2\sigma^2 + \nu^2, \tag{A.20}$$

respectively. In (A.19), $_1F_1(\cdot; \cdot; \cdot)$ represents the generalized hypergeometric function. Further details can be found in [PM06].

## A.3  Nakagami-$m$ Distribution

The Rayleigh and Rician distributions can be employed to model the statistics of some physical properties of the channel models, such as the fading envelope. Nevertheless, these distributions do not always provide an accurate fit to measured data.

The Nakagami-$m$ distribution is also employed to characterize the statistics of signals transmitted through channels with multipath fading channels. In fact, the Nakagami-$m$ distribution frequently provides a closer fit to experimental data than the Rayleigh or the Rician distribution [PM06]. The PDF of the Nakagami-$m$ distribution is given by [PM06]

$$f_R(r) = \begin{cases} \frac{2}{\Gamma(m)} \left(\frac{m}{\Omega}\right)^m r^{2m-1} e^{-\frac{mr^2}{\Omega}}, & m \geq 1/2, r \geq 0, \\ 0 & r < 0, \end{cases} \tag{A.21}$$

where $\Gamma(\cdot)$ is the Gamma function, $\Omega$ denotes the second moment of the random variable $R$ given by

$$\Omega = \mathbf{E}(R^2), \tag{A.22}$$

and the parameter $m$ is the Nakagami shape factor fading parameter which ranges from $1/2$ to $\infty$, and is defined as the ratio of the moments [PM06]

$$m = \frac{\Omega^2}{\mathbf{E}\{(R^2 - \Omega)^2\}}, m \geq 1/2, \tag{A.23}$$

allowing the Nakagami-$m$ distribution to correspond to several of the multipath distributions. Let us consider, for example, the special cases: when $m = 1/2$ the Nakagami-$m$ fading channel corresponds to the one-sided Gaussian distribution; when $m = 1$ it corresponds to the Rayleigh distribution, and when $m \to \infty$ it converges to a non-fading AWGN channel.

The $k^{th}$ moment of $R$ is given by

$$\mathbf{E}[R^k] = \frac{\Gamma(m + \frac{k}{2})}{\Gamma(m)} \left(\frac{\Omega}{m}\right)^{\frac{k}{2}}, \tag{A.24}$$

and the variance by

$$\text{Var}[R] = \Omega \left[1 - \frac{1}{m} \left(\frac{\Gamma(m + \frac{1}{2})}{\Gamma(m)}\right)^2\right]. \tag{A.25}$$

The Nakagami-$m$ consists of a sort of general fading distribution whose parameters were defined so that they could be adjusted to adapt to different empirical measures. Moreover, its PDF is also known to frequently provide closed-form solutions in system performance studies.

# Appendix B

# Complex Baseband Representation

Since the message bearing signal $s(t)$ is physically realizable, it consists of a real-valued bandpass signal, and consequently the corresponding spectrum $S(f)$ is centered in a carrier frequency $f_c$, and symmetric around 0 Hz, as depicted in Fig. B.1. So, we may say that $S(f) = S(-f)$. Since the signal

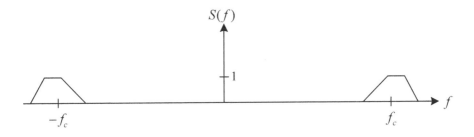

**Figure B.1: The spectrum $S(f)$.**

is real, the signal's $s(t)$ information is localized in the positive part of the spectrum $S(f)$ which can be represented by $S^+(f) = 2S(f)U(f)$, where $U(f)$

is the unit step function given by

$$U(f) = \begin{cases} 0 & f < 0 \\ 1/2 & f = 0 \\ 1 & f > 0 \end{cases} \tag{B.1}$$

An example of $S^+(f)$ is illustrated in Fig. B.2. The inverse Fourier transform

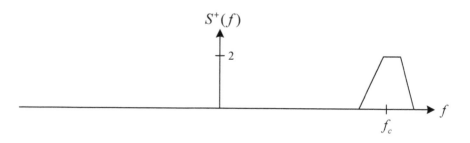

**Figure B.2: The spectrum $S^+(f)$.**

of $S^+(f)$ can be given by

$$\begin{aligned} s^+(t) &= \mathcal{F}^{-1}\{S^+(f)\} \\ &= \mathcal{F}^{-1}\{2S(f)U(f)\} \\ &= \mathcal{F}^{-1}\{2U(f)\} * \{S(f)\} \end{aligned} \tag{B.2}$$

Applying the inverse Fourier transform of $\mathcal{F}^{-1}$ to B.2

$$\mathcal{F}^{-1}\{2U(f)\} = \delta(t) + j\frac{1}{\pi t} \tag{B.3}$$

we may write

$$\begin{aligned} s^+(t) &= \left(\delta(t) + j\frac{1}{\pi t}\right) * s(t) \\ &= s(t) + j\frac{1}{\pi t} * s(t) \end{aligned} \tag{B.4}$$

A simplification of equation (B.4) can be made with resort to the Hilbert transform of $s(t)$, by making

$$\mathcal{H}\{s(t)\} = \frac{1}{\pi t} * s(t) = \hat{s}(t) \tag{B.5}$$

and equation (B.4) can be rewritten as

$$s^+(t) = s(t) + j\hat{s}(t) \tag{B.6}$$

From (B.6) it becomes clear that if we pass the signal $s(t)$ through a linear system with an impulse response given by $h(t) = \frac{1}{\pi t}$, it will result in signal $s^+(t)$.

The frequency response of the linear system can be obtained by performing the Fourier transform of the impulse response $h(t)$,

$$H(f) = \mathcal{F}\{h(t)\} = \begin{cases} j & f < 0, \\ 0 & f = 0, \\ -j & f > 0. \end{cases} \tag{B.7}$$

Therefore for the spectrum $\hat{S}(f)$, results

$$\hat{S}(f) = H(f)S(f). \tag{B.8}$$

Let us now denote the equivalent baseband signal by $s_b(t)$. The signal $s_b(t)$ can be obtained from $s^+(t)$ with resort to a frequency translation of its spectrum; this is

$$S_b(f) = \frac{1}{\sqrt{2}}S^+(f + f_c) \tag{B.9}$$

where $\frac{1}{\sqrt{2}}$ is a scaling factor, and $f_c$ the translation frequency (i.e., the carrier frequency). An example of $S_b(f)$ is illustrated in Fig. B.3.

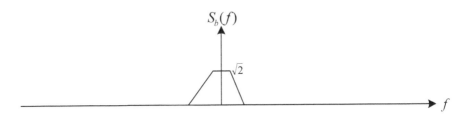

**Figure B.3: Equivalent baseband signal spectrum.**

From $S_b(f)$ we can obtain the equivalent baseband signal $s_b(t)$ (which is also known as the complex envelope of $s(t)$), with resort to the inverse Fourier transform,

$$s_b(t) = \mathcal{F}^{-1}\{S_b(f)\} = \frac{1}{\sqrt{2}}s^+(t)e^{-j2\pi f_c t}. \tag{B.10}$$

Applying the result of (B.6) to (B.10), we can rewrite it as follows

$$s_b(t) = \frac{1}{\sqrt{2}} \left( s(t) + j\hat{s}(t) \right) e^{-j2\pi f_c t}. \tag{B.11}$$

If we rewrite (B.11) as

$$s(t) + j\hat{s}(t) = \sqrt{2} s_b(t) e^{j2\pi f_c t} \tag{B.12}$$

and by knowing that $s(t)$ and $\hat{s}(t)$ are real signals, it is clear that $s(t)$ can be obtained from $s_b(t)$ by taking the real part of (B.12), given by

$$s(t) = Re\left\{ \sqrt{2} s_b(t) e^{j2\pi f_c t} \right\}. \tag{B.13}$$

The complex baseband representation (or complex envelope) $s_b(t)$ can be written in terms of its real and imaginary parts as

$$s_b(t) = s_{I(t)} + j s_{Q(t)} \tag{B.14}$$

From this, (B.13), and applying the Euler's identity we get

$$s(t) = \sqrt{2} \left[ s_{I(t)} \cos(2\pi f_c t) - s_{Q(t)} \sin(2\pi f_c t) \right]. \tag{B.15}$$

The complex baseband representation can also be represented in polar form. If we define the envelope $a(t)$ and phase $\psi(t)$ as follows,

$$a(t) = |s_b(t)| \sqrt{s_{I(t)}^2 + s_{I(t)}^2}, \tag{B.16}$$

and

$$\psi(t) = \tan^{-1} \frac{s_{Q(t)}}{s_{I(t)}}, \tag{B.17}$$

then we get

$$s_b(t) = a(t) e^{j\psi(t)}. \tag{B.18}$$

If we apply the above equations to (B.14), we get

$$s_b(t) = \sqrt{2} a(t) \cos\left[ 2\pi f_c t + \psi(t) \right] \tag{B.19}$$

This notation, known as baseband-passband representation, is often used to model the wireless signal transmission. Before the transmission the baseband signal is upconverted to the chosen carrier frequency, at the transmitter side. At the receiver side the received signal is downconverted back to the baseband.

# Appendix C

# Minimum Error Variance

In Chapter 5 we proposed a channel estimation method based on training sequences multiplexed with data. It was shown that it is possible to use a decision-directed channel estimation to improve the accuracy of channel estimates without requiring high-power training sequences. Here we show how we can combine the channel estimates, obtained from the training sequence, $\tilde{H}_k^{TS}$, with the decision-directed channel estimates, $\tilde{H}_k^D$, to provide the normalized channel estimates with minimum error variance defined in (5.11).

Let us assume the channel estimates,

$$\tilde{H}_k^D = H_k + \epsilon_k^D, \tag{C.1}$$

and

$$\tilde{H}_k^{TS} = H_k + \epsilon_k^{TS}, \tag{C.2}$$

where the channel estimation errors, $\epsilon_k^D$ and $\epsilon_k^{TS}$, are assumed to be uncorrelated, zero-mean, Gaussian random variables with variance $\sigma_D^2$, and $\sigma_{TS}^2$, respectively, i.e., $\epsilon_k^D \sim N(0, \sigma_D^2)$ and $\epsilon_k^{TS} \sim N(0, \sigma_{TS}^2)$. The channel estimates $\tilde{H}_k^D$, and $\tilde{H}_k^{TS}$, can be combined as follows:

$$\tilde{H}_k^{TS,D} = \frac{a\tilde{H}_k^D + b\tilde{H}_k^{TS}}{a+b} = \frac{\tilde{H}_k^D + \dfrac{b}{a}\tilde{H}_k^{TS}}{1 + \dfrac{b}{a}} = \frac{\tilde{H}_k^D + \mu\tilde{H}_k^{TS}}{1 + \mu} = H_k + \epsilon_k^{TS,D}, \tag{C.3}$$

where $a = b = 1$, $\mu = \dfrac{a}{b}$, and $\epsilon_k^{TS,D} \sim N(0, \sigma^2)$ denotes the noise component, still characterized by a Gaussian-distribution, with zero mean and variance $\sigma^2$, given by

$$\sigma^2 = \frac{\sigma_D^2 + \mu^2 \sigma_{TS}^2}{(1 + \mu)^2} = f(\mu). \tag{C.4}$$

For the sake of simplicity, we dropped the dependence with $k$. The parameter $\mu$ is chosen to minimize $\sigma^2$. The optimum value of $\mu$ corresponds to

$$\frac{df(\mu)}{d\mu} = 0, \tag{C.5}$$

leading to

$$\mu = \frac{\sigma_D^2}{\sigma_{TS}^2}. \tag{C.6}$$

Therefore the overall channel estimate combining, resulting from the combination between $\tilde{H}_k^{TS}$ and $\tilde{H}_k^{D}$, will be

$$\tilde{H}_k^{TS,D} = \frac{\sigma_{TS}^2 \tilde{H}_k^D + \sigma_D^2 \tilde{H}_k^{TS}}{\sigma_D^2 + \sigma_{TS}^2} = H_k + \epsilon_k^{TS,D}, \tag{C.7}$$

where $\epsilon_k^{TS,D} \sim N(0, \sigma_{opt}^2)$ denotes the noise component with Gaussian-distribution, with zero mean and variance $\sigma_{opt}^2$. The optimum variance $\sigma_{opt}^2$ will be

$$\sigma_{opt}^2 = \sigma^2 \Bigg|_{\mu = \frac{\sigma_D^2}{\sigma_{TS}^2}} = \frac{\sigma_D^2 + \left(\frac{\sigma_D}{\sigma_{TS}}\right)^4 \sigma_{TS}^2}{\left(1 + \frac{\sigma_D^2}{\sigma_{TS}^2}\right)^2} = \frac{\sigma_D^2 \sigma_{TS}^4 + \sigma_D^4 \sigma_{TS}^2}{(\sigma_D^2 + \sigma_{TS}^2)^2} = \frac{\sigma_D^2 \sigma_{TS}^2}{\sigma_D^2 + \sigma_{TS}^2}. \tag{C.8}$$

Under these conditions results, $\sigma_{opt}^2 \leq \sigma_D^2$ and $\sigma_{opt}^2 \leq \sigma_{TS}^2$.

# References

[AD04]      T. Araújo and R. Dinis.   Iterative equalization and carrier
            synchronization for single-carrier transmission over severe time-
            dispersive channels. *IEEE GLOBECOM'04*, pages 204–212, 2004.

[BDFT10]    N. Benvenuto, R. Dinis, D. Falconer, and S. Tomasin. Single Car-
            rier Modulation with Nonlinear Frequency Domain Equalization:
            An Idea Whose Time Has Come — Again. *Proceedings of the
            IEEE*, 98(1):69–96, 2010.

[Bin90]     J. A. C. Bingham. Multicarrier modulation for data transmission:
            an idea whose time has come. *IEEE Communications Magazine*,
            28(5):5–14, 1990.

[BT02]      N. Benvenuto and S. Tomasin.  Block iterative DFE for single
            carrier modulation. *Electronics Letters*, 38(19):1144–1145, 2002.

[BT05]      N. Benvenuto and S. Tomasin. Iterative design and detection of
            a DFE in the frequency domain. *IEEE Transactions on Commu-
            nications*, 53(11):1867–1875, 2005.

[CH03]      C. Cozzo and B. Hughes.  Joint channel estimation and data
            detection in space-time communications. *IEEE Transactions on
            Communications*, 51(8):1266–1270, Aug 2003.

[Cha66]     Robert W. Chang. Synthesis of band-limited orthogonal signals
            for multichannel data transmission. *The Bell System Technical
            Journal*, 45(10):1775–1796, 1966.

[Chu72]     D. Chu.  Polyphase codes with good periodic correlation prop-
            erties (corresp.).  *IEEE Transactions on Information Theory*,
            18(4):531–532, Jul 1972.

[Cim85] L. Cimini. Analysis and simulation of a digital mobile channel using orthogonal frequency division multiplexing. *IEEE Transactions on Communications*, 33(7):131–141, 1985.

[CKB06] Y. Choi, C. Kim, and S. Bahk. Flexible design of frequency reuse factor in ofdma cellular networks. In *ICC '06. IEEE International Conference on Communications*, volume 4, pages 1784–1788, June 2006.

[Cla98] M. Clark. Adaptive frequency-domain equalization and diversity combining for broadband wireless communications. *IEEE Journal on Selected Areas in Communications*, 16(8), 1998.

[CT65] James W. Cooley and John W. Tukey. An algorithm for the machine calculation of complex Fourier series. *Mathematics of Computation*, pages 297–301, 1965.

[DAPN10] R. Dinis, T. Araújo, P. Pedrosa, and F. Nunes. Joint turbo equalisation and carrier synchronisation for SC-FDE schemes. *European Transactions on Telecommunications*, 21(2):131–141, 2010.

[DB93] P. Dent and G. Bottomley. Jakes fading model revisited. *Electronics Letters*, 29(13):1162–1163, June 1993.

[DG04] R. Dinis and A. Gusmo. A class of nonlinear signal processing schemes for bandwidth-efficient OFDM transmission with low envelope fluctuation. *IEEE Transactions on Communications*, 52(11), 2004.

[DGE01] L. Deneire, B. Gyselinckx, and M. Engels. Training sequence versus cyclic prefix-a new look on single carrier communication. *Communications Letters, IEEE*, 5(7):292–294, July 2001.

[DGE03] R. Dinis, A. Gusmo, and N. Esteves. On broadband block transmission over strongly frequency-selective fading channels. *15th International Conference on Wireless Communications*, pages 261–269, 2003.

[DKFB04] R. Dinis, R. Kalbasi, D. Falconer, and A. Banihashemi. Channel estimation for MIMO systems employing single-carrier modulations with iterative frequency-domain equalization. *Vehicular Technology Conference, 2004 IEEE*, 7:4942–4946, 2004.

[DLF04] R. Dinis, C. Lam, and D. Falconer. On the impact of phase noise and frequency offsets in block transmission CDMA schemes. *IEEE ISWCS'04*, pages 131–141, 2004.

[DLF07] R. Dinis, C. Lam, and D. Falconer. Joint frequency-domain equalization and channel estimation using superimposed pilots. *IEEE WCNC'08*, pages 90–96, 2007.

[DLF08]      R. Dinis, Chan-Tong Lam, and D. Falconer. Joint frequency-domain equalization and channel estimation using superimposed pilots. In *Wireless Communications and Networking Conference, 2008. WCNC 2008. IEEE*, pages 447–452, March 2008.

[DMCG12]     V. Dalakas, P. Mathiopoulos, F. Cecca, and G. Gallinaro. A comparative study between SC-FDMA and OFDMA schemes for satellite uplinks. *Transactions on Broadcasting*, 58(3):370–378, 2012.

[FABSE02]    D. Falconer, S. Ariyavisitakul, A. Benyamin-Seeyar, and B. Eidson. Frequency domain equalization for single-carrier broadband wireless systems. *Communications Magazine*, 4(4):58–66, 2002.

[For73]      G. D. Forney. The viterbi algorithm. *Proceedings of the IEEE*, 61(3):268–278, 1973.

[GCG79]      J. Gain, G. Clark, and J. Geist. Punctured convolutional codes of rate (n-1)/n and simplified maximum likelihood decoding. *IEEE Transactions on Communications*, 25:97–100, 1979.

[GDCE00]     A. Gusmo, R. Dinis, J. Conceio, and N. Esteves. Comparison of two modulation choices for broadband wireless communications. *IEEE Vehicular Technology Conference*, 2:1300–1305, 2000.

[GDE03]      A. Gusmo, R. Dinis, and N. Esteves. On frequency-domain equalization and diversity combining for broadband wireless communications. *IEEE Transactions on Communications*, 51(7):1029–1033, 2003.

[GTDE07]     A. Gusmo, P. Torres, R. Dinis, and N. Esteves. A turbo FDE Technique for reduced-CP SC-based block transmission systems. *IEEE Transactions on Communications*, 55(1):16–20, 2007.

[HKR97]      P. Hoher, S. Kaiser, and P. Robertson. Pilot-symbol-aided channel estimation in time and frequency. *IEEE GLOBECOM'97*, pages 90–96, 1997.

[JC94]       William Jakes and Donald Cox, editors. *Microwave Mobile Communications*. Wiley-IEEE Press, 1994.

[JCWY10]     T. Jiang, H.-H. Chen, H.-C. Wu, and Y. Yi. Channel modeling and inter-carrier interference analysis for vehicle-to-vehicle communication systems in frequency-dispersive channels. *Mobile Networks and Applications*, 15(1):4–12, 2010.

[Kai95]      S. Kaiser. On the performance of different detection techniques for ofdm cdma in fading channels. *IEEE Globecom 95*, 3:2059–2063, 1995.

[LFDL08] Chan-Tong Lam, D. Falconer, and F. Danilo-Lemoine. Iterative frequency domain channel estimation for dft-precoded ofdm systems using in-band pilots. *Selected Areas in Communications, IEEE Journal on*, 26(2):348–358, February 2008.

[LFDLD06] Chan-Tong Lam, D. Falconer, F. Danilo-Lemoine, and R. Dinis. Channel estimation for sc-fde systems using frequency domain multiplexed pilots. In *Vehicular Technology Conference, 2006. VTC-2006 Fall. 2006 IEEE 64th*, pages 1–5, Sept 2006.

[Lin95] F. Ling. Matched filter-bound for time-discrete multipath Rayleigh fading channels. *IEEE Transactions on Communications*, 43:710–713, 1995.

[LMWA02] M. Loncar, R. Muller, J. Wehinger, and T. Abe. Iterative joint detection, decoding, and channel estimation for dual antenna arrays in frequency selective fading. In *The 5th International Symposium on Wireless Personal Multimedia Communications*, volume 1, pages 125–129, Oct 2002.

[LSW81] R. Lucky, J. Salz, and E. Weldon. *Principles of Data Communication*. McGraw-Hill, 1981.

[MAG98] S. Merchan, A. G. Armada, and J. L. Garcia. OFDM performance in amplifier nonlinearity. *Transactions on Broadcasting*, 44:106–114, 1998.

[Mat05] A. Mattsson. Single frequency networks in DTV. *Transactions on Broadcasting*, 51(4):413–422, 2005.

[MM96] V. Mignone and A. Morello. CD3-OFDM: a novel demodulation scheme for fixed and mobile receivers. *IEEE Transactions on Communications*, 1996.

[MM99] M. Morelli and U. Mengali. An improved frequency offset estimator for OFDM applications. *IEEE Communications Letters*, 3:75–77, 1999.

[Moo94] P-H. Moose. A technique for orthogonal frequency division multiplexing frequency offset correction. *IEEE Transactions on Communications*, 42(10):2908–2914, 1994.

[MR93] F. Muller-Romer. DAB-the future European radio system. *EE Colloquium on Terrestrial DAB — Where is it Going?*, 33:1–6, 1993.

[NF04] B. Ng and D. Falconer. A novel frequency domain equalization method for single-carrier wireless transmissions over doubly-selective fading channels. *IEEE GLOBECOM'04*, 2004.

[Pat03]     Matthias Patzold. *Mobile Fading Channels.* John Wiley & Sons, Inc., 2003.

[PDN10]     P. Pedrosa, R. Dinis, and F. Nunes. Iterative frequency domain equalization and carrier synchronization for multi-resolution constellations. *Transactions on Broadcasting*, 56(4):551–557, 2010.

[PM06]      John G. Proakis and Dimitris K. Manolakis. *Digital Signal Processing.* Prentice-Hall, Inc., 4th edition, 2006.

[Rap01]     Theodore Rappaport. *Wireless Communications: Principles and Practice.* Prentice Hall PTR, 2nd edition, 2001.

[Rei95]     U. Reimers. The DVB project-digital television for Europe. *IEE Colloquium on DVB (Digital Video Broadcasting): The Future for Television Broadcasting?*, 33:1–7, 1995.

[Ric44]     S.O. Rice. Mathematical analysis of random noise. *The Bell System Technical Journal*, 23(3):282–332, July 1944.

[SC97]      T. Schmidl and D. Cox. Robust frequency and timing synchronization for OFDM. *IEEE Transactions on Communications*, 45:1613–1621, 1997.

[SDM10]     F. Silva, R. Dinis, and P. Montezuma. Joint detection and channel estimation for block transmission schemes. *Military Communications Conference (MILCOM)*, pages 1765–1770, 2010.

[SF08]      M. Sabbaghian and D. Falconer. Joint turbo frequency domain equalization and carrier synchronization. *IEEE Transactions on Communications*, 7(11):204–212, 2008.

[SKJ94]     H. Sari, G. Karam, and I. Jeanclaude. An analysis of orthogonal frequency division multiplexing for mobile radio applications. *IEEE Vehicular Technology Conference*, 3:1635–1639, 1994.

[Skl97]     B. Sklar. Rayleigh fading channels in mobile digital communication systems characterization. *IEEE Communications Magazine*, 35(7):90–100, 1997.

[TH00]      M. Tuchler and J. Hagenauer. Turbo equalization using frequency domain equalizers. *Allerton Conference*, 2000.

[TH01]      M. Tuchler and J. Hagenauer. Linear time and frequency domain turbo equalization. *IEEE VTC 2001*, 2001.

[VY02]      B. Vucetic and J. Yuan. *Turbo codes: Principles and Applications.* Kluwer Academic Publ., 2002.

[WE71]    S. Weinstein and P. Ebert. Data transmission by frequency-division multiplexing using the discrete fourier transform. *IEEE Transactions on Communication Technology*, 19(5):628–634, 1971.

[WHW08]   H. C. Wu, X. Huang, and X. Wang. Theoretical studies and efficient algorithm of blind ICI equalization for OFDM. *Transactions on Wireless Communications*, 7(10):3791–3798, 2008.

[WPW08]   H.-C. Wu, J. Principe, and X. Wang. Robust switching blind equalizer for wireless cognitive receivers. *IEEE Transactions on Wireless Communications*, 7(5):1461–1465, 2008.

[WW05a]   H. C. Wu and Y. Wu. A new ICI matrices estimation scheme using Hadamard sequence for OFDM systems. *Transactions on Broadcasting*, 51(3):305–314, 2005.

[WW05b]   H. C. Wu and Y. Wu. Distributive pilot arrangement based on modified m-sequences for OFDM intercarrier interference estimation. *Transactions on Wireless Communications*, 51(3):1605–1609, 2005.

[WWC⁺09]  X. Wang, H.-C. Wu, S. Y. Chang, Y. Wu, and J.-Y. Chouinard. Efficient non-pilot-aided channel length estimation for digital broadcasting receivers. *Transactions on Broadcasting*, 55(3):633–641, 2009.

[WY12]    H.-C. Wu and Y. Yan. Novel robust BPE-IWLMS blind equalizer for phase shift-keying signals. *Transactions on Communications*, 60(11):3174–3180, 2012.

[WYWS10]  Z. Wang, Z. Yang, J. Wang, and J. Song. Frequency domain decision feedback equalization for uplink SC-FDMA. *Transactions on Broadcasting*, 56(2):253–257, 2010.

[XW05]    Songnan Xi and Hsiao-Chun Wu. Fast channel estimation using maximum-length shift-register sequences. In *Vehicular Technology Conference, 2005. VTC-2005-Fall. 2005 IEEE 62nd*, volume 3, pages 1897–1900, Sept 2005.

# Index